45
5

Taming The Taildragger

by

JOHN BALL

A Flight Manual

For

Classic Tailwheel Type Aircraft

4th Edition 1990

FOREWORD

The idea for this manual was prompted by the lack of a single source of information about flying older,"taildragger" type aircraft. While operating a flight school using an Aeronca Champ, I wrote a few pages of pertinent techniques and procedures I wanted available to pilots transitioning to tail-wheel planes for the first time. Over the years I have expanded it to include new tips from "oldtimers", firsthand information from owning nine different tailwheel types, and answers to the most common questions I've been asked. I think you will find a lot of practical information packed between these covers.

Flying these classic, conventional airplanes captures flying as it used to be - simple, relaxing, and fun. I hope this manual will add to your pleasure.

John Ball

Although the author and publisher have researched many sources to ensure the accuracy and completeness of information contained in this book, we assume no responsibility for errors, inaccuracies, omissions, or any other inconsistency herein. This book is intended to be used in conjunction with instruction from a Certified Flight Instructor (CFI).

ISBN 0-929214-00-5 PAPERBACK

0-929214-02-1 HARDCOVER

TABLE OF CONTENTS

STARTING

Starting an aircraft engine by pulling the propeller through by hand, called "propping", has become a lost art around most modern day airports. Most pilot's think of it as a last resort for getting an engine started after the battery has gone dead. As the only means of starting many of the older classic taildraggers and some homebuilt aircraft, we should look at the procedures required to "prop" a plane successfully and safely.

Although the safety record of people propping aircraft in recent times is poor, it is the result of lack of knowledge or disregard of a few simple precautions. The two most common problems are the engine starting unexpectedly, causing the propeller to strike the person propping, and the aircraft leaping forward because the throttle was too far open during the start. This often results in the aircraft taxiing into something or even becoming airborne with no one aboard !

If at all possible, never attempt to prop a plane without a KNOWLEDGEABLE person in the cockpit. Be sure this person knows the signals or terms you will use, where all the required controls are and, if necessary, how to shut the engine off.

For the rare times it is not possible to have someone in the cockpit, at least insure that the aircraft is securely tied down until the engine is started and idling slowly. The "Maintenance" section of this manual shows a handy tiedown line that may prove useful if you find yourself with no one to help you start.

Here are two interesting ways bush pilots sometimes use when starting engines without help: One is to shut off the fuel valve prior to propping the engine after priming and filling the carburetor. If all goes well there will be time to turn it back on after start before the carburetor runs dry. If the aircraft gets away from him, it will run out of gas before it gets very far. Another idea is to tie a line to the throttle or mixture handle and to the pilot's wrist so that if the plane starts to taxi, the line will pull the throttle or mixture off even if the pilot slips and can't get to the cockpit in time to stop it. These are extreme measures and should only be used as a last resort.

Check the ground for solid footing where you will be standing while propping. Wet grass, loose gravel, or snow/ice should be avoided whenever possible. Floatplanes are usually started by the person propping the engine standing on the righthand float and pulling the prop from the rear.

Treat all propellers as though the mags are hot. Aircraft ignition systems are wired so that a broken wire or bad connection will result in the ignition being on. As a result, even though the mag switch may be off the engine might start if the propeller is pulled.

When ready to start, check that the fuel shutoff valve is open, open the throttle about 1/4 inch, and insure the mag switches are off. If the plane is equipped with a mixture control, it should be in the "Full Rich" position. Many of the older classic planes with small engines have no mixture control - they are full rich all the time.

If necessary, use the primer pump. Be careful not to overprime as this will allow excess fuel to collect in the intake manifold and possibly cause a fire during start. There is no accelerator pump on most of the smaller engines so pumping the throttle does not prime them. To spread the priming fuel through the engine, pull the prop through compression six to eight times.

Call for brakes on and switches hot ("Contact" if you are a traditionalist!) The person in the cockpit repeats these calls as he does them. To insure the brakes are holding, push on the prop hub. Grasp the prop with the fingertips over the top of the blade and the thumb on the front side - NOT hooked under it. Where you place your hands along the length of the blade will be determined by the size of the engine being started. On large engines that require a large force to pull through compression you will need as much leverage as possible so you should pull from the very tip of the prop. On small, easily propped engines, placing your hands half way between the tip and hub will cause the engine to be spun faster than if it is grasped at the end. This will give a hotter spark, especially if the engine is not equipped with an impulse mag. Impulse mags are those that have a mechanism to spin the mag faster than the engine would turn it normally. They can be recognized by the "clunk" sound heard even when the prop is turned over slowly.

Practice moving your body away from the propeller's path as you pull it through. Then, should you slip you will fall away from the whirling prop rather than into it. Ideally, the prop will come against engine compression when it is at the "10 o'clock" position. This the safest for most aircraft as you will not be off balance when pulling the hardest. If yours does not, an A&P can rotate it on its hub until it does. Low propellers require extra caution as the person pulling the prop through is in an awkward, bent over position. Incidentally, engines with large or metal props are often easier to start than those with small or wooden ones as the inertia of the heavier prop will help carry it through more firing strokes than the lighter prop. When propping a large engine, for example a Continental 220, it will help if you back the prop to the compression force before pulling it through. The prop inertia will help overcome the compression resistance of the next cylinder and thus help you get more plugs to fire for each pull. Lifting your right leg and swinging it down and back as you pull the prop will help give more force to your pull as well as help get the momentum of your body to pull you away from the prop arc. For small engines this is not necessary.

As soon as the engine starts, insure that the throttle is at idle and check the oil pressure. If there is no oil pressure within thirty seconds, shut the engine down and investigate. Check that the level is not too low in the tank. Sometimes, especially on small engines, the oil pump may have lost its prime. Picking up the tail for a few seconds either before or after the engine is running will often restore the prime. Oil that is too light or a worn pump will make this problem worse.

Hot engine starts should be attempted without the preliminary pull-throughs. If the engine fires once, but kicks back, it is probably too rich. If propping it once more does not start it, it may be flooded. With the throttle full open and the mags off, pull the prop backwards 3 to 4 revolutions, and try another start after retarding the throttle back to idle. Pulling the prop backwards does not clear the engine any faster than pulling it forward, but there is less chance of the engine accidentally firing.

Another way to start an engine that has been flooded is with the throttle wide open. This is VERY dangerous and can only be accomplished with a competent person in the cockpit to quickly retard the throttle once the engine starts. Failure to do so has resulted in aircraft running over the person propping it, ramming nearby aircraft, or even getting airborne inadvertently!

Let me repeat: To safely prop an engine requires one to resist complacency, use common sense, and follow a few simple precautions.

TAXIING

After start, the engine should be idled at approximately 1000 RPM until oil temperature comes off the peg. This can be accomplished while taxiing on most planes.

When taxiing over dirt or gravel, don't use more power than necessary to keep moving as this may cause the propwash to pick up stones and nick the propeller. Also keep carburetor heat off as it bypasses the air filter and allows dust to be sucked into the engine.

Most tailwheel aircraft have tailwheel steering and this should be used as much as possible. For aircraft without this feature it will be necessary to use differential braking for steering. Even steering equipped aircraft will require differential braking for sharp turns. Although you shouldn't carry it to an extreme, taxiing fast enough to maintain forward momentum will assist your turn when differential braking is required. Applying, then releasing the brake before you come to a complete stop is more effective than locking up a wheel, coming to a stop, and then using a large blast of power to pull the plane around. If you have come to a stop and need to turn sharply, keep the rudder deflected, open the throttle slowly, and CAUTIOUSLY ease the stick forward to take some weight off the tailwheel. Be very careful when applying forward stick so you don't lift the tail off the ground and risk striking the prop. In some airplanes with heel brakes it may be necessary to use one foot to hold full rudder and the other foot to press the brake simultaneously.

In no wind or light wind conditions, you should keep the stick aft to keep weight on the tailwheel for steering and to keep from nosing over if you hit soft ground or a depression in the field. Propwash will help press the tail down even without help from the natural wind.

When taxiing in windy conditions it will be necessary to position both the elevator and ailerons properly to retain control. When taxiing with the wind from the rear, the stick should be forward - with the wind from the front, it should be back. The idea is to use the elevator to press the tailwheel onto the ground. If you are uncertain whether the wind is strong enough from the rear to keep the stick forward probably you should keep it aft due to the effect of the propwash.

With a crosswind quartering from the front, place the stick back and toward the side from which the wind is coming. Cross-winds from the rear quarters require the stick to be positioned forward and on the opposite side from the wind. This will help keep the aircraft level and not allow the wind to get underneath a wing and lift it up. These control positions are the same for tricycle and tailwheel airplanes but are probably more important for the latter because they are so light, have large control surface area, and are often high winged. As a result it is easier for the wind to affect them.

Turns while taxiing downwind should be as slow as possible because of the weathervane tendency. Use the ailerons as necessary in such turns to keep the upwind wing down.

If you have a choice of direction when making a turn, you can often use the weathervane tendency to your advantage. If you turn into the wind, the wind will assist the start of the turn and, as the aircraft aligns with the wind, it will slow the turn. This will help prevent you from overshooting the heading. On the other hand, if you turn away from the crosswind, you will be hindered initially, then when the tail passes directly by the wind, it will swing rapidly until you are headed into the wind. Turn in the direction that will be the most controllable.

Another condition that may affect your taxi technique is runway "crown" or slope. Many runways are built with the center higher than the sides to assist drainage. When taxiing on these runways, stay to the upwind side so that the tilt of the plane will help keep the upwind wing down. Once the wind "gets under" a lightly loaded wing it is easy for it to lift it high enough to put the aircraft out of control.

Some tailwheel planes have restricted forward visibility. If yours does, taxi in an "S" pattern to clear the area ahead of you. Sometimes leaning out of an open window or door helps, too. Don't forget to look out for soft spots and depressions in the taxiway.

This section wasn't meant to scare you about the difficulties of taxiing light, tailwheel aircraft, but keep in mind that you do have to be especially careful in windy conditions. If you get caught in very high winds and you must move the aircraft, try to get some wing walkers and maybe someone to hold the tail down while you taxi.

Prior to takeoff, the aircraft should be turned into the wind, the brakes held, the stick held all the way back, and the engine runup to the proper RPM. Do this over a hard surface free of stones or sand if possible to protect the propeller. Some original owner's manuals called for full static power RPM. By today's standards this seems excessive unless you are checking full power for maintenance reasons. Usually an RPM of 1700 to 2000 is used. Whatever you use, always use the same one for that particular plane. This will help you detect any change in the ignition system from flight to flight.

Check the magnetos by switching from "Both" to "Left", back to "Both", "Right", and back again to "Both". Maximum allowable RPM drop varies with aircraft type - see operating numbers listed in the back of this manual for those of some common types. More important than the size of the drop is any tendency for rough running on one mag.

Be suspicious of a zero RPM drop. It may be due to a magneto that is "hot" all the time because of a switch or lead wire malfunction. Occasionally you will notice a larger drop on the left mag than the right. On some engines the left mag operates the bottom sparkplugs while the right mag fires the top ones. The bottom ones are more apt to foul and thus the difference.

Sometimes an engine that runs rough during its runup can be cleared by running it at full RPM and then leaning it, if it has a mixture control. If carbon buildup (often caused by 100 octane avgas) was the reason, this may burn it off and allow normal operation. Don't do this for more than minute or so as you may overheat your engine badly.

After the mags are checked, pull out the carburetor heat knob to check for an RPM drop. This check is misleading on some aircraft as it shows little drop on the ground and yet it is working properly and will show a large drop in the air. You should learn what is normal for your plane and let that be your guide.

It is not necessary to wait until the engine temperatures reach normal levels. If the engine will accelerate to full power smoothly it is warm enough for takeoff. Some manufacturers listed minimum oil or cylinder head temperatures.

Cycle the flight controls through their full range to check for proper movement and any restrictions. Set the trim where your experience has shown is best. Tandem aircraft require a larger difference in trim settings than side by side ones. You might make two marks to show where the trim should be for solo and with a passenger. Be prepared to hold whatever forces are necessary to maintain the proper attitude if you are flying an aircraft for the first time and there is no takeoff setting marked.

If possible you should clear the entire pattern for aircraft before taking the runway. A good way of doing this in a high winged airplane is to turn a full circle in place.

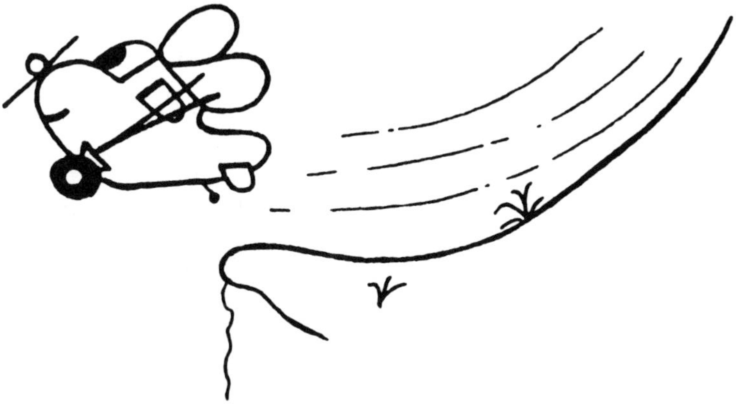

TAKEOFF

After lining up on the center of the runway, hold the stick full back and smoothly advance the throttle to full power. Quickly verify that the normal takeoff RPM is reached. If operating from a dirt runway, ease the throttle forward slowly to allow the aircraft to accelerate a little before full power is reached. This will minimize the number of rocks pulled up into the propeller.

As the tail begins to feel light, place the stick slightly forward of neutral to raise the tail. If you raise the tail too rapidly, the plane is apt to swerve to the left due to the gyroscopic effect of the engine and propeller. With the tail up the visibility will improve greatly.

As speed increases the rudder will become more effective to control the heading so smaller and smaller rudder pedal deflections are needed. Making small corrections early is better than waiting until large ones are needed. This will minimize "S"-ing during the takeoff roll. Most pilots prefer to leave their heels on the floor and press the rudder pedals with their toes.

At liftoff speed ease the stick back and the aircraft will lift into the air. Keep the climb shallow until the climb airspeed is reached. Increase the climb attitude to maintain this speed. With the small engines installed in most of these small aircraft, you should use full throttle until reaching cruise altitude.

If you are leaving the pattern after takeoff turn crosswind for a few seconds, then turn 45 degrees away from the field to indicate to other pilots your intention. Do not make any turns until you are at least 200 feet above the ground.

Counter the tendency for the plane to turn left by using right rudder. Few of these planes have a cockpit rudder trim so they are set up to fly without any rudder pedal deflection at cruise speed and power. As a result they will require rudder pressure during all climbs. Many european engines, including Volkswagen powered ones, rotate counterclockwise so require left rudder.

Make clearing turns while climbing so you can check for other planes that might be hidden by the nose. Pilot visibility is very poor in many of these planes - especially those that are soloed from the rear. You will have to check the area that you are turning into BEFORE you roll into a bank. With a high wing and low windows it is nearly impossible to see that area once you are in the bank.

CRUISE

After leveling off at the desired altitude, allow the aircraft to accelerate to cruise speed at full power, taking care not to exceed the maximum allowable RPM. After reaching the cruise speed throttle back to the RPM that will maintain it.

Proper cruise RPM may not be known due to the variety of propeller pitches authorized for a type of plane/engine. To find the approximate RPM, some manufacturers recommended using the same RPM as maximum static RPM found on the ground. Another method used by some is to fly at full throttle as close to sea level as possible and use 90% of that RPM at cruise.

Always use full throttle when climbing to a new altitude. Engine manufacturers recommend it as there is less time spent at the higher power setting and the cooling is better because the mixture is enrichened at full throttle.

If most of your flying time is spent at cruise, you should adjust your rudder trim to allow cruising without holding a rudder pedal deflected. On most of the older planes this means bending the tab on the rudder trial and error until you find the correct setting. Bend it a little at a time as it is usually quite effective. Bend it in the direction opposite the rudder pedal you had to hold at cruise (e.g. if right rudder pedal held, bend the tab to the left).

If most of your flying is done in a climb (such as glider towing) bend the tab to ease the forces on the rudder during climb.

Do not blindly trust the manufacturer's cruise performance charts. Most, especially the older ones, are very optimistic. They were based on no wind conditions, a new engine, a properly rigged aircraft, no allowance for climb and no reserve fuel at the destination. If you combine this with the lack of accuracy of many fuel and airspeed indicators, you can see how a real problem might develop. Other pilots and your own experience can supply more accurate information.

If it is necessary to cruise with carburetor heat on, allow for a richer mixture and thus an increased fuel consumption.

The benefits in speed and range that flying at high altitudes usually brings will not be available on aircraft not equipped with a mixture control. The air/fuel mixture gets richer with decreasing air density. If the mixture cannot be leaned, the benefits of flying at a more efficient airspeed at altitude will be cancelled by a too rich mixture. It is only worthwhile to climb several thousand feet if there will be a sizable tailwind there.

PATTERN

When descending after using high power settings, do not pull the throttle all the way to idle immediately. Allow the engine to cool down at approximately 1500 RPM for a minute or so. This will be much easier on the engine due to slower cooling.

Use carburetor heat anytime the RPM is below 2000 as these engines (especially Continental 65's) are prone to carb ice. This can happen at surprisingly high air temperatures if the humidity is high.

Whenever you are near an airport be especially alert for the midair collision hazard. Most of the planes we are discussing in this book are high winged ones so it will be necessary to clear the airspace in the direction of your intended turn before you roll into it. When you spot another plane, you can determine if you are on a collision course with it or not by his drift across the windscreen or window. If he appears to drift forward, you will pass behind him. If he drifts backward, you will pass in front of him. If he appears stationary - Look Out!- you are on a collision course. If you make a turn to avoid him, make it obvious so that he won't turn in a direction (trying to avoid you) that will continue the hazard.

When landing at a strange airport with no radio, fly over it at 1500 feet above the ground to check the windsock, traffic, and runway condition. Lefthand turns are standard. If it is a completely unknown field and it looks little used, fly a low pass along the full length to check for wet spots, tall grass, ruts, and obstructions such as ditches.

A hazard common to many small airstrips is the location of power lines across the ends, right in your approach or departure path. Watch for poles near the end of any airstrip as they are much easier to see than the wires. If a road runs by one end of the strip there is a greater chance that wires will be there.

You should be at the correct altitude prior to entering the pattern. Most airports today use 1000 AGL as pattern altitude. Some airports prefer 800 feet. The proper one is listed in the Airman's Information Manual or other airport directory. A 600 foot pattern is more comfortable for many of the older low powered aircraft, but should only be used at private strips or when no one else is in the pattern.

When on downwind, opposite the touchdown point, pull the carb heat on, throttle back to approximately 1500 RPM and hold 10 MPH above final approach speed in the glide. When the runway is about 45 degrees back from the wingtip, turn to base. At this point check carefully for anyone making a straight-in approach and adjust your power as needed to obtain the correct final approach speed and glide angle on final.

Instructors once insisted that all patterns be flown at idle power from downwind to touchdown. This makes every landing a practice "dead stick" landing. Unfortunately, this will usually place you well inside everyone else's pattern on base so, if you do this today, be very watchful for planes on longer finals making powered approaches. If you make a normal "square" pattern and carefully check the final approach area when on base you can avoid such a conflict.

LANDING

This section is probably the reason most people buy this manual. It is the landing phase of flight that most separates the tail-wheel plane from the tricycle geared one. We will cover basic techniques and emphasize the most important points, but it will take practice to become really proficient.

For your first practice sessions, try to find an instructor familiar with tailwheel planes to teach you. This may not be as easy as it sounds as many instructors have no experience with them either. At least find someone with experience to demonstrate some landings for you. You might try a local EAA chapter for possible help. They are usually friendly and are happy to see someone interested in basic aircraft.

The best type of field to practice on is a grass airstrip. Small errors in alignment will not have as large an effect as on a hard surface runway.

Don't complicate your first few practice sessions with crosswinds or gusty winds. Keep it simple and as your proficiency grows, you can take on more challenging wind conditions. If there is a question in your mind whether to fly or not - don't.

Don't be scared away by hangar stories of groundloops and bounces. They can't be too bad - these were the primary trainers of a few years ago. Modern trainers tend to be forgiving of imprecise (dare I say "sloppy"?) flying. Sometimes proper speed control, rudder use and glidepath control are sacrificed for radio procedures and operation of aircraft systems. With these rather simple planes, flown from uncrowded airstrips, one has the time to concentrate on accurate aircraft control.

If there is one "secret" to good taildragger landings it is consistent, accurate patterns that put you at the end of the runway at the correct speed, at the right altitude, and with proper alignment so that you can concentrate on the actual flare and landing. So let's look closely at the final approach.

When turning from base to final be very careful not to use too much rudder. There is a common tendency for some pilots to "hurry" the turn with rudder, especially if they are overshooting final. Too much rudder will cause the plane to overbank. When the pilot opposes this with aileron control away from the turn he is applying prospin controls. Some planes are more susceptible to this because they have light rudder forces. Just remember to keep all your turns coordinated and you will avoid these problems.

As soon as possible after turning final, line up with the center of the runway. This minimizes "angling" finals which require last minute turns close to the ground. Use throttle and stick to maintain final approach speed and the correct angle of descent. Tailwheel aircraft are more sensitive to variations in landing speeds than tricycle geared ones so you must fly more precisely.

To assist in landing on a desired spot, pick an "aimpoint" 40 to 50 feet in front of it. By watching that point drift up or down the windscreen you can tell if you are maintaining the correct glidepath. If the spot drifts down the windscreen you are above the glidepath. If it drifts up you are below it. If you choose the spot correctly, keep it close to one spot on the windscreen, and maintain the proper speed, your flare will take you to the desired touchdown point.

With a few exceptions, flaps on these older planes are not very effective (if they are available at all). If necessary to steepen your approach and you have no flaps, you should slip the aircraft. It is usually a good practice to add 5 mph while slipping to allow for airspeed indicator errors and to insure that you won't stall in this cross controlled condition.

Realign the aircraft with the runway at about 30 feet of altitude unless you need the wing low for a crosswind. It is better to concentrate on airspeed control and touchdown alignment from that point on. If you allow an excessive airspeed to build, you will float a long way in these lightly wingloaded planes before it dissipates. Pushing the nose over to avoid floating will probably cause the plane to bounce back into the air.

At 15 to 20 feet above the ground, begin to slow your rate of descent. Be careful not to level off this high. Let the airspeed begin to decay as the plane settles into ground effect. As the aircraft slows, bring the stick further and further back to keep the sink rate from increasing rapidly. Use the throttle as necessary to control airspeed/sink rate, too. Attempt to hold the main wheels a foot or two off the runway as long as possible. The goal is to touch down at minimum speed and full aft stick with all three wheels touching at the same time.

Sometimes, particularly with someone in the back seat of a tandem plane, the tail-wheel will touch first. This doesn't hurt anything if it isn't from too great a height. You can be sure the plane will not bounce if this is the case.

To determine the proper three point landing attitude, you should notice where the horizon intersects the cowling as you taxi out before takeoff. Then, when "feeling for the runway", you will be able to recognize the approximate landing attitude to hold.

Once the flare has begun, looking two to three aircraft lengths ahead and to the sides will allow you to judge your height after a little practice. Be careful not to fix your gaze on the horizon or too close to the aircraft as it will be difficult to make consistent landings that way. If you sense a rapid sink rate beginning and you are in about the right landing attitude, ease the throttle forward to arrest it before you touchdown. Don't hesitate to go around and try again if things begin to get out of hand.

Even the best pilot will occasionally bounce a landing. Often this arises when he tries to land too fast and, when the wheels do touch, he "flinches" which pulls the plane back into the air. If he then pushes the nose down rapidly he will probably start the cycle all over again. This is an example of a PIO which stands for "pilot induced oscillation". If caught soon enough, before the bounces get too large, it may be possible to complete the landing. The pilot should hold the aircraft in the proper landing attitude, let the speed bleed off to normal, and then bring the stick back as in a normal landing. If the first bounce is very large, or the next ones increase in size, the best procedure is to hold the stick slightly aft, open the throttle, and go around. The landing gear on these planes can withstand quite large bounces without damage as long as the plane stays aligned with its flightpath. They are not as strong when subjected to sideloads.

If you go around, you should be aware that many of these planes have a powerful trim control. You can usually trim for stable, "hands off" flight on final approach. However, keep in mind that the strong noseup trim that was just right while gliding at idle power will pitch the nose up rapidly when you add power for the go around. This is due to the propwash over the elevator and the increasing airspeed You can either retrim for the climb condition or hold the stick pressures necessary to maintain the correct climb attitude.

To "wheel land" a tailwheel plane, fly a flatter than normal approach with partial power on. Once the rate of descent is slowed, fly lower and lower until the main wheels touch. Quickly pin the wheels to the runway by moving the stick forward an inch or two. Slowly decrease power until the tail settles to the runway. The landing roll will be longer than normal because of the faster touchdown speed and a slower decrease in power. As you will still have flying speed at touchdown, if you hurry the tail lowering you may get airborne again. Take your time. On some airplanes the airflow over the rudder will be blanked by the fuselage as the tail comes down, but the situation won't last long.

Use as much or more care to maintain directional control during the landing roll-out as you did during the takeoff. It is a little more difficult during landing because there is no propwash flowing past the rudder to increase its effectiveness. Any sideways drifting at touchdown can result in a large lateral oscillation or even a ground loop.

DO NOT RELAX after landing until the aircraft has slowed to taxi speed. Once the speed is low enough that the aircraft can not become airborne again, bring the stick full back to increase pressure on the tailwheel for better steering.

Be very careful if using brakes for directional control. Probably more damage has been done by the overuse of brakes than by the lack of their use. Some larger tailwheel planes require braking for control because the tail is blanked out by the fuselage in the three point attitude. This not true in most small planes as the rudder is effective enough. Often the brakes on these older planes are not very reliable anyway. You will probably be better off to let the plane just roll to a stop. This won't take much distance because of the slow landing speed.

Tailwheel planes seem to be happiest on grass fields! They look more at home there and they are certainly easier to land there. This is at least partially due to the slipperiness of grass that allows a small drift at touchdown without imparting a large sideload on the landing gear that might result in the beginning of a groundloop. Also there is a small amount of cushioning by the grass that makes most landings softer.

Run the engine at 800 to 900 RPM prior to shutting it down. Occasionally the engines that do not have a mixture control to use for stopping the engine will continue to run (very roughly) after the magnetoes are switched off. This is called "dieseling" and can usually be remedied by opening the throttle as the engine coasts to a stop. Another possibility is to run on each mag for about 30 seconds to help cool the sparkplugs faster.

If dieseling is not a problem, you can sometimes charge the cylinders with fuel for the next start by leaving the throttle closed during shutdown. Then, if you plan on starting again within an hour or so, it won't be necessary to prime the engine.

If the engine continues to run smoothly after the mag switch is off, there probably is a broken wire or bad connection in the magswitch circuit. The only method left to shut the engine down if there is no mixture control is to shut off the fuel valve and wait for the carburetor to run out of gas.

TIEDOWN

When tying a plane down you should position the anchors so that they form a triangle like this :

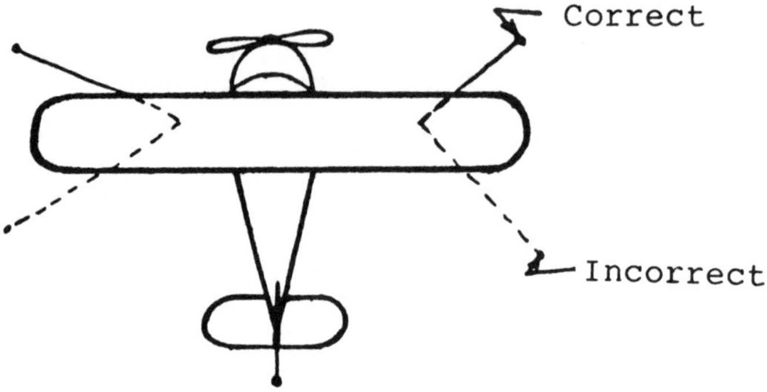

If using ropes to tie the plane down, leave enough slack to allow for shrinking if they become wet and then dry out. Tie the tail down with a rope through the lift handle or around the tailwheel support. This rope should not go around the steering springs but should pass between them and the tailwheel support.

An excellent means of anchoring tiedown ropes to the ground is to use dog chain anchors available at hardware stores for a few dollars. They are the same anchors as sold in aviation supply stores for much more.

Some manufacturers recommend lowering full flaps when the plane is tied down outside. Usually these are the planes with flaps that are held up by springs so lowering them provides a more positive means of preventing them from blowing in the wind and experiencing excessive wear.

Most aircraft have some kind of tiedown ring under their wings. On older aircraft these are notoriously weak from corrosion or fatigue. A method that doesn't rely on the strength of these rings on aircraft with lift struts is to tie them down this way :

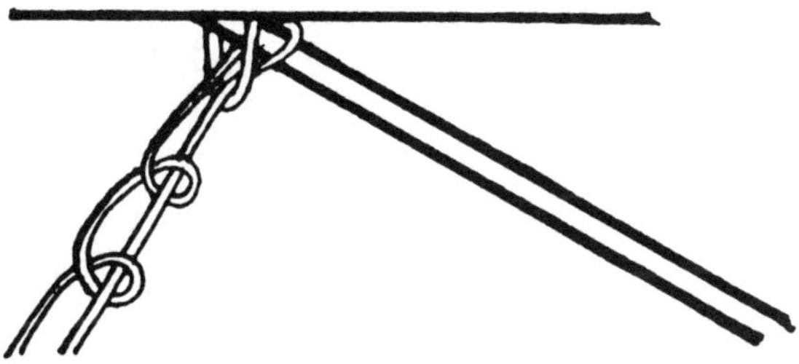

One disadvantage of tailwheel planes is their tail low attitude during high winds. If winds are coming from their front their wings have a rather high angle of attack and will exert a large lift force on the tiedown ropes. One method of minimizing this is to tie them down with their main wheels in a ditch or other depression. You can also tie them down with their tails toward the expected winds. Use caution though because, if the winds are very strong, the wings can see a large downward force that may cause damage.

A walk down most tiedown areas will show you that most airplanes are left with their controls held aft, usually with the seat-belt. Although this is better than letting them flop in the wind, they really should be tied forward. The elevator will drain better when it is down because usually the drain holes are in the trailing edge. If the plane has tandem seating, the rear stick can be tied forward with the front seatbelt. If a plane has control wheels instead of sticks it will require gustlocks. See the maintenance section of this manual for more information.

CROSSWIND

Ground handling in windy conditions was covered in the earlier section on taxiing. Just remember to place the controls correctly, taxi on the upwind side of a crowned runway, and make all turns into the wind if possible.

For a crosswind takeoff, keep the tail down until the rudder is effective enough to keep the fuselage aligned with the runway. At the beginning of the roll, use full aileron into the wind to keep the upwind wing down. As the speed increases less aileron is needed so reduce the deflection.

Keeping the plane aligned with the runway with the rudder, allow the plane to lift off with the upwind wing low. Use a higher take-off speed (+5 mph) than normal to insure the aircraft won't touch down again. Once you are positively airborne turn a few degrees toward the wind to counteract the drift.

Because of the low speeds usually involved during takeoff, landing, and even cruise, winds will have a larger effect than on faster planes you may be familiar with. Some basics of flying that are often forgotten in more modern planes suddenly take on a new importance when applied to older and slower planes. The downwind leg should be moved closer to the runway if the wind is coming from the direction of that runway and moved farther out if coming from the opposite direction.

Adjusting the lead points as necessary, roll out on final aligned with the runway as soon as possible. Most pilots use the wing low method to keep from drifting downwind. Of course opposite rudder is needed to keep from turning. Monitor your airspeed closely as being crosscontrolled close to the ground can mean real trouble should you stall. Add one half the expected wind gust to your final approach speed. Wind strength will usually decrease close to the ground because the air slows down when it flows around trees, buildings, etc.

Most tailwheel planes should be wheel landed in crosswinds. The upwind main wheel should touch first. As speed dissipates, lower the other main and finally the tailwheel. Hold aft stick to press it down for positive steering. By using a flatter than normal approach with power on and landing a little faster than normal you will keep the rudder more effective in controlling your heading. Land on the upwind side of a crowned runway to help keep that wing down.

A good practice maneuver for crosswind landing is to fly down final with your wing low to kill the drift. Then, when you are at flare height, add power to hold a foot or two off the runway and fly the whole length of the runway that way. You will be able to see how good a correction you have very easily.

If the crosswind is definitely too strong, consider landing at another airfield that is aligned more with the wind. If none are within range, even a road or pasture may be a safer landing area. If the wind is so strong that you can't land as originally planned, the landing roll should be very short if you turn directly into it. In an emergency, some pilots have even landed across a normal runway or on a taxiway.

SHORT FIELDS

To takeoff from a short field, especially one with obstacles at the end (don't they all ?), use every foot of the runway available. Pull the plane backwards by hand if necessary to get it to the very end.

Apply power smoothly, but without hesitation, to full throttle. Some pilots prefer to run up to full power before releasing brakes. Others think the propeller won't "bite" the air until there is forward motion. Use whichever technique you prefer, but be careful of high power settings over loose gravel or you may do considerable damage to your prop.

During the initial roll use neutral control to minimize aerodynamic drag. After the tail has flown up, don't allow the aircraft to liftoff until just prior to best angle of climb speed. Apply back pressure to the stick to maintain best angle of climb speed until over any obstacles, then lower the nose to attain normal climb speed. If you have them, half flaps just before liftoff may shorten the takeoff roll considerably.

For a short field landing, fly a normal pattern except hold 5 - 10 MPH slower than normal on final. Use full flaps on final if you have them. If it appears the glide angle is too steep, don't pull the nose up to stretch the glide. Add power to flatten the glide angle. Don't push the the stick forward after passing an obstacle on the approach end in an effort to get the plane on the ground as soon as possible. This is the natural tendency, but will increase your speed and cause the plane to float for some time before touching down.

If you maintain the proper final approach speed there is very little floating after the flare. A small burst of power may help arrest an excessive sink rate just before touchdown.

Make a real attempt to land at the slowest safe speed, probably tailwheel first. You should have full aft stick at touchdown so just leave it there and apply brakes carefully and evenly. You will be surprised at how short a landing roll most of these planes have if the proper technique is used.

Be very careful using this procedure in gusty wind conditions as you will be close to stall and a sudden gust or quick control movement may cause you to settle into the obstacle.

When forced to decide between taking off downhill with a tailwind or uphill with a headwind, most pilots prefer the former unless the tailwind is very large.

Don't forget to consider the runway surface. Mud, tall grass, or snow can extend your takeoff roll further than you might think.

If you are contemplating pushing your aircraft's limits, take all factors into account - wind, temperature, runway slope, obstacles, type of runway surface, etc. It may be necessary to lighten the gross weight by leaving some fuel, passengers, or baggage behind.

Most of these planes were never blessed with excess power so density altitude affects their performance drastically. Conditions may require postponing the flight until more favorable winds or temperatures arrive. If in doubt, don't go. It's not worth it.

SOFT FIELDS

The basic idea of soft field takeoffs is to get the weight off the wheels as soon as possible because rolling drag will be greater than the aerodynamic drag if the field is soft or snow covered. To accomplish this, pick the tail up as soon as possible. After the tailwheel is off the ground ease the stick aft as quickly as you can without letting the tailwheel touch the ground again. This will cause the plane to get airborne at a very slow speed due to ground effect.

As speed increases stay in this ground effect by moving the stick forward. This will help the plane accelerate to normal climb speed. Trying to climb out of ground effect too soon may prevent you from accelerating for a long time.

Fly a normal pattern when landing on a soft field. Hold the aircraft off the runway as long as possible to touch down at minimum speed. The stick should be nearly full aft to minimize the chance of nosing over. The touchdown speed won't need to be as slow as for a short field, but almost.

If the mainwheels sink into mud during the rollout, their drag will go up dramatically and may be enough to cause the plane to nose over. Holding yourself back from the instrument panel with your left arm may keep you from going forward, pushing the stick forward and helping the noseover tendency. Brakes probably won't be needed if the field is very soft. In fact you may have to taxi a little faster than normal with power to keep from bogging down. Try not to stop until you are where you want to be or are on solid ground as it is often more difficult to start moving again than it is to keep the wheels rolling.

It should be emphasized that these techniques, especially the takeoff, can be dangerous in gusty wind conditions.

These techniques can also be used on rough fields where prolonged ground roll may damage the landing gear.

COLD WEATHER

The first suggestion is : DRESS WARM to prevent the freezing off of a vital part of your anatomy ! The older planes (especially the fabric ones) were equipped with only minimal cabin heaters at best. Many are also drafty so, unless you are right next to the heater opening, what little heat they do produce will be lost before it reaches you. The back seat of a tandem aircraft is particularly chilly. If you are planning a lot of flying in cold weather you should look into auxiliary heaters available for some aircraft.

If your plane is not hangared you will have to insure that all the snow is off your wings before attempting to fly. Contrary to what you might expect, snow does not always blow off during takeoff. If it is loose snow there will be no problem pushing it off with a broom. If there is a layer of ice under the snow it may be very difficult to scrape that off without damaging the wing. If possible try to pull it into a heated hangar after clearing all loose snow off or leave it so the sun will strike it at the best angle to melt it off. One way or another, make sure the wing is clear before taking off as a surprisingly thin layer of snow or ice can affect lift drastically.

One other item to check on preflight is the venting of tanks - fuel and oil. The oil breather tube is susceptible to freezing closed due to the moisture produced as the oil is warmed. A hole drilled on the back side of the tube keeps some air available for venting even if the end is frozen shut.

Even more important is the fuel vent. If it gets completely plugged with snow or ice, enough of a vacuum may be created in the tank to stop fuel flow to the engine! A good preflight inspection should eliminate both problems.

The next problem is getting the engine started. You should have the proper oil in your engine for the temperature. See the lists of aircraft specifications at the back of this manual for suggestions. If the oil is too heavy it will hinder the turning of the engine by the starter (which may be your arm if you are propping it). Many pilots have been using multi-weight oil with good results.

Even with the proper oil, the engine will be more difficult to rotate than normal. You should help free the engine by turning it over several revolutions by hand even if it has a starter so that it will cause the least resistance when attempting to start. Remember to treat it as though the mags are hot any-time you are moving the prop.

If the outside temperature is below 15 to 20 degree F, you should consider preheating the engine with either an aviation preheater or a substitute you can supply. Some types that have been used before include : a 100 watt lightbulb left under the cowling for a few hours; a small space heater sold in most discount stores; a duct taped to the dis-charge opening of a car heater and led to the heater and led to the cowling; and a personal hairdryer. Although some people have used the exhaust from a car as a preheater, it contains so much moisture and carbon that it should be considered only as a last resort.

Placing blankets, coats, etc over the cowling will maximize the effects of any heater but will be useless if just left over a cold engine overnight with no other source of heat.

If propping the plane because the battery is dead or the aircraft doesn't have a starter, be careful of your footing on the snow covered ground. Some pilots prefer to prop from behind the propeller. This gives them the advantage of being able to hold onto the lift strut for support and , should the aircraft slip on the snow after the engine starts, it will move away from the person rather than toward him.

More priming fuel will be needed in cold weather. On the small engines such as 65 HP there is no acceleration pump on the carburetor so pumping the throttle will not supply any priming fuel. The priming pump should be used and the engine pulled through to spread the fuel throughout the cylinders before turning the mags on and attempting a start in the usual manner.

On starter equipped engines the battery may not be able to supply enough power to crank a very cold engine. You can help it by "jumpstarting" with your automobile battery (if both are the same voltage) or you can install an external power receptacle so a larger battery can be used.

Another way to help your starter in cold weather is, after pulling the propeller through several times to loosen the oil, move the prop backwards until it comes up against compression. This will allow the engine to turn without the load of compression for a few degrees. The propeller will then have some momentum to carry it through the compression stroke. Every little bit helps.

If the engine fires a few times and then refuses to fire any more, the sparkplugs may have "frosted". This means that enough moisture was generated with the few firings that frost has formed across the sparkplug electrodes. Once this has happened the only cure is to warm the engine to melt the frost or remove the plugs to clear them.

After the engine is beginning to run, it may be necessary to prime it more to keep it running. Pumping the throttle may be sufficient with some engines, but on the smaller ones you will have to operate the primer pump carefully until the engine can run smoothly without it. Be certain that you lock the primer pump closed as, if it opens in flight, it can dump enough fuel into the engine to cause it to run poorly and possibly quit.

Carburetor heat will sometimes allow the engine to warm up and run smoother during taxi. Usually grass and dust are frozen during very cold weather so will not be pulled into the engine even though the carb heat bypasses the airfilter.

It is not necessary to wait a long time for all temperatures to reach their normal levels. If the oil pressure is not reading too high and the engine will smoothly accept full throttle, the engine is warm enough for takeoff.

"Cold air kits" consisting of aluminum plates to cover air inlets and cloth covers for intake manifolds and oil sumps are available for some planes. These help the engine warm up and slow cooldown times.

The actual flying in cold weather is practically the same as in warm weather. If there is snow on the runway, use soft field techniques for takeoff and landing. It will help to keep some power on during the approach so the engine will be warm in case you need to go around. This will also make the carburetor heat more effective.

If you live in an area that has a constant snow cover for weeks at a time, you may want to install skis on your plane. These tailwheel planes are ideal for this because of their low takeoff and landing speeds, prop clearance, and they don't need a third ski.

The initial installation should be done by an A&P. He can calculate the change in CG and make any required modifications. After this initial time, the owner/operator can change back and forth between wheels and skis by himself. Some installations retain the wheel and tire as a shock absorber -- others bolt directly to the axle. Usually the tips are held up slightly by bungee cords while the plane is airborne.

Although there is no special license or rating required to fly ski-equipped aircraft, it would be a good idea to fly with a pilot familiar with ski operations before trying it on your own.

A plane on skis flies about the same as when equipped with wheels. It may cruise a little slower due to the extra drag of supporting cables and the misalignment of the skis with the airstream.

Landings and takeoffs are usually done using basic soft field procedures. This minmizes the jarring of the airframe. The type of snow (texture,weight,depth) affects takeoff and landing distances considerably. Deep, loose snow will lengthen takeoff distances, but will shorten the landing distance. Crosswinds may pose a problem with these slow landing planes, especially on ice encrusted snow. Determining snow conditions while airborne is often difficult because it all looks the same and a few degrees of temperature can drastically affect the characteristics.

When landing on an open, snow-covered field, it may be difficult to judge your height above the ground. In this case, maintain a gradual, flatter-than-normal descent, with some power left on, until touchdown. This technique will require a longer distance to stop, but that is better than stalling while several feet in the air.

Place something, such as a board, underneath the skis when tying a skiplane down. This will keep them from freezing to the snow. A fast taxi will generate enough heat to melt snow when you stop. If you don't do something to keep the skis from freezing to the snow you will be stuck. Hot water may be used to thaw them out, but it will be necessary to taxi immediately or the skis will just refreeze to the snow.

CROSS COUNTRY

Some of the most enjoyable cross country flying one can do can be done in an older taildragger at 70 MPH. At the common altitude of 1000 feet AGL one sees scenery missed by both the motorist and the pilot roaring along at 10,000 feet centering the needle between VOR stations. Enjoying the scenery, meeting interesting people at the many little airstrips where you'll be stopping, and just generally taking life easy are excellent therapy for the "hustle mania" of modern life.

Many pilots are not comfortable with the thought of flying long distances without the help of IFR instruments and radios. Actually, these planes can be flown safely (and legally) almost anywhere. Long trips are just a series of short ones added together !

This section will discuss some topics you should consider before starting out. With a little common sense and practice you will soon learn how much fun "getting back to basics" can be.

Special Equipment:

Some form of ear protection is almost mandatory for extended flights. It is amazing how tiring the constant high noise level can be - not to mention the ear damage that can result. Although ear plugs are preferred by some, many pilots prefer the "Mickey Mouse"type ear protectors worn by skeet shooters.

Other handy items you might include are a watch (very few older planes had built-in clocks), a chamois in case you must fill your tank from cans, and an airport directory to help you plan overnight stops.

There must be some barnstorming spirit in all pilots. Many like the feeling of "roughing it" by sleeping in a sleeping bag underneath the wing. Not all airport operators will let you do it however. In any case carrying a sleeping bag is not a bad idea if you have the room.

Finally, a basic tool kit plus safety wire, cloth duct tape, and parachute cord or other strong line may prove useful.

Navigation:

For a pilot who has been navigating by the modern method of VOR "needle centering", the prospect of flying cross country with nothing but a map and compass creates some apprehension. Brushing up on some of those long forgotten techniques of map reading will help ease these concerns. Some pilots try to go too far with the "keep it simple" theme" and use road maps. In today's world of restricted areas, TCA's, and large airport traffic areas, this is not a good idea. You will need more information than just which road goes where.

Because of the low ground speeds involved, sectional maps are the most useful. Usually your navigation problem consists of following a large highway, railroad or power line. Highways offer the advantage of supplying a potential emergency landing site. Major thruways usually have a clear grassy area between traffic lanes that might prove handy sometime. Even though it is often illegal to use them, in an emergency that would be of little concern. Road signs may be a salvation for the "temporarily disoriented" pilot.

Railroads and power lines are worth considering for navigation because they often run the shortest distance between two cities. Keep a close watch on your compass when following them as not all are shown on maps and it is easy to follow the wrong one !

Winds will affect these slower planes more than they would faster ones. The difference of only 20 MPH in groundspeed will change your range by over 25 % if you are flying a standard J3 Cub for example. Knowing the actual winds becomes very important when one considers the short range of most older taildraggers.

Even crosswinds while cruising decrease your range more than in faster planes. The crab angle you must hold to maintain a certain track across the ground is larger. This reduces your speed in the direction of your destination.

You may have to change your destination while airborne if the winds affect you enough. Always have a field picked out closer than your original destination if the headwind turns out to be stronger than planned.

To help you determine your groundspeed and thus your range while flying, draw a line along your route on your map before taking off. Rather than a series of straight legs, you may find it easier to draw the line along the road you expect to follow. Mark it off every ten miles. A mark scratched on a pencil corresponding to ten miles on your map will speed this chore. When at cruise speed, check your groundspeed between any two prominent points rather than between checkpoints 30 minutes or more apart. Calculating your speed in miles/minute is often easier than doing it in MPH.

Simple mathmatics can be used to determine time remaining to your destination. Simply time how long it takes to go 10 miles (from one tick mark on the map to the next) and multiply by the number of 10 mile segments left to the destination.

Try to find the best altitude to maximize tailwinds or minimize headwinds. Forecasted winds are notoriously inaccurate, especially at the low altitudes we are talking about. Surface winds can be estimated from ripples in water or grain, the waving of flags or clothes on a line (not too common in most parts of the country today!), and the drifting of smoke. The winds at the cloud level can be estimated from the speed and direction of their shadows along the ground. If you know the winds at these two levels you will have help in determining how high to fly.

You can get help increasing your groundspeed by using winds other than simple tailwinds. If the wind is blowing perpendicular to a ridge, you can get additional lift by flying on the upwind slope of that ridge. Gliders use the effect of the air flowing up the face of the slope to climb. You can use it to increase your speed by, in effect, diving at the same rate the air is rising. Often this will result in an increase of 10 to 15 MPH. Be careful not to get downwind from the ridge though as it is usually very turbulent there and the air will be descending, causing you to raise the nose to maintain altitude and thus slow down.

There will be times when it will be necessary to navigate by dead reckoning. Assuming you understand the fundamentals, here is an idea that will help you find the correct heading to hold without knowing the winds beforehand. Prior to starting the dead reckoning leg, find two points (towers, road intersections, etc.) that are aligned with your next checkpoint. Fly directly from one point to the next noting the heading needed to maintain the desired track across the ground. Assume the wind will remain the same and simply hold that heading to your next checkpoint.

If possible choose a checkpoint that is located on a road, railroad, river, etc that runs perpendicular to your flightpath. This makes it difficult to miss the point as far as distance is concerned. A problem might occur if you arrive at the road or whatever and don't know whether you are left or the right of the checkpoint. To prevent this, you might hold a heading 5 to 10 degrees left or right to insure knowing which way to turn in case you don't hit the checkpoint exactly.

If visibility allows, use any prominent landmark such as a mountain peak past your checkpoint to keep lined up with your course. You can use this method even if turbulence makes your compass oscillate badly. This is usually easier in the clear skies of the western US than in the murk and haze of the East.

Remember, you are flying for the fun of it. Worrying about whether you have enough fuel or not takes the fun away very fast. If in doubt - land !

These suggestions may seem simple, but I have used them to complete crosscountry trips of over 3,000 miles.

Weather:

The greatest obstacle to cross country flying in this type of plane is the requirement for VFR weather. By carefully choosing your route and airports, it is legal to "scud run" (fly low under clouds) with a ceiling of 500 feet and one mile visibility. A more practical minimum however is 1000 feet and 3 miles. This will keep you legal almost everywhere and is much safer.

If the weather closes in, do a 180 degree turn to get out of it or find a friendly looking field and land. The danger of landing in an unknown field may be less than continuing into deteriorating weather.

Keep a close watch for carburetor icing. Many pilots think that because the outside air temperature is above freezing, ice can't form in the carburetor. This is not true due to the cooling effect of the evaporation of the fuel and the decrease in pressure in the venturi of the carburetor. The humidity level is as good an indicator of icing probability as temperature. A damp or muggy day can cause a carburetor to ice up surprisingly fast. Icing has been reported in temperatures as high as 80 degrees Fahrenheit.

To help detect carb icing, set cruise power on a day when you are sure there is no icing. Mark that throttle position on a nearby stationary surface so that you can align the throttle with that mark later. Always set your cruise power by using the mark. Then, should the RPM's drop below the normal cruise with the throttle set correctly, don't adjust the throttle - pull the carb heat on. Pushing the carb heat back off after a couple of minutes will probably bring the engine RPM back to normal and verify that you had icing.

If it is necessary to fly with the carb heat on continuously (not very common), remember it will make the mixture rich and shorten your range. If your engine has a mixture control you should lean it if prolonged use of carb heat is expected.

Although the chances of a rain shower drowning out the ignition are slim, you should avoid flying in showers that you can't see through. The loss of visibility may be more dangerous than the water.

The desire to stay out of a solid shower caused a landing on the Alaska Highway during my first trip to Alaska. Rather than divert around a small storm located directly on the highway, I chose to find a straight section of road, land, and wait for it to pass. Most motorists gave little if any notice to my plane parked miles from any airport !

Radio:

Contrary to what many pilots trained in the last couple of decades may believe, a radio is not essential to safe cross country flying. By knowing the proper procedures and enough regulations to stay legal, you can go almost anywhere in the country without any kind of radio. There are many times more airports without towers than with.

All tower equipped airports are surrounded by an "airport traffic area" extending in a five mile radius around the center of the field and up to 3000 feet AGL. With a few exceptions, you should not enter that airspace without contacting the tower.

One exception occurs if you are planning to land or takeoff from an "uncontrolled" airport located within that boundary. Of course if you have a radio it is a good idea to talk to the tower at the controlled field to inform them of your intentions, but it is not required.

If your route of flight takes you too close to a field with a tower, flying at least 3000 feet AGL will keep you above its traffic area and it will not be necessary to contact them.

Larger airports often have "Terminal Radar Service Areas" (TRSA) around them. Use of this advisory service is voluntary so you would be legal to fly through it (avoiding the five mile airport traffic area mentioned above). It would be advisable to avoid this if possible due to the density of traffic. In any case, be very watchful for other airplanes when near any airport.

If you are not familiar with "Terminal Control Areas" (TCA's) around very large airports, such as Chicago, Atlanta, Los Angeles, etc you should visit an FAA Flight Service Station or a large FBO to check the Airman's Information Manual (AIM) for information. Any flight instructor should be able to explain them to you also.

Many of the classic tailwheel planes have antiquated radios. These are often powered with only a battery or a wind driven generator mounted in the slipstream. Some of them have very few frequencies, are heavy, and are unreliable. Often the cost to repair them is more than they are worth. Many are illegal because of frequency "spillover".

If you are contemplating the purchase of a radio remember to add the cost of the antenna, microphone, speaker, shielded ignition, shielded sparkplugs, and electrical power supply you may need. There are portable, handheld comm radios available today that are equipped with a built-in antenna, microphone, speaker, and rechargeable battery. They have from 10 to 720 frequencies. The plane must still have shielded ignition wire and sparkplugs though.

If you want a VOR capability also, you can buy a simple NavCom that can be powered by a rechargeable battery pack or aircraft electrical system. These are quite cheap, as aircraft radios go, and are sufficient for the type of VFR flying we are discussing.

Although they can prove useful sometimes, radios somehow seem to dilute the pure fun of flying these older taildraggers.

EMERGENCIES

These aircraft are simple and reliable so few malfunctions occur. In fact the emergency section of one owner's manual written in 1947 consisted solely of the statement, "Standard emergency procedures apply." That is beyond the imagination in today's world! This section should cover most of these "standard emergency procedures".

Engine Problems:

Most people are overly concerned about engine failures. Modern engines are reliable and rarely stop without some prior warning. In your entire flying career you may never experience one failure if you are careful. However being rare is not the same as being impossible so we should not become complacent and should be aware of the potential problems.

Fuel Contamination - The most common contamination is water. Keeping the fuel tank full when the plane is idle for long will prevent condensation from forming in the tank. Some tanks are sensitive to deterioration of the rubber seal around the cap.

A useful item for planes that use the cork/wire type gauge for indicating fuel quantity (eg J3 Cub) is a plastic cover from an aerosol can. By placing it over the fuel cap, the wire will be held down inside the tank and will seal the cap from rain when the plane is tied down outside. If not protected, the wire often rusts and hangs up on the cap. By keeping it immersed in gasoline, it is prevented from rusting and contaminating the fuel. You should be able to find a cover that fits your cap tightly.

Rust, sand, and other foreign particles are not very common and should be caught during periodic fuel filter inspections. Refueling from a recently filled underground tank may cause enough foreign particles to be suspended in the fuel to clog fuel filters and even carburetor jets.

If it is necessary to use metal cans to fuel your plane, use a chamois skin to filter the fuel. This is the biggest worry for many pilots who are using automobile gas in their engines.

If the plane has been idle for quite a while, drain a considerable amount of fuel from the tanks and gascolator into a glass container during preflight. Tests have shown that water may not show up until several quarts have been drained. If the container is clean and no contamination is found, you can put it back into the tank.

Raise the tail to a level flying attitude when draining fuel if you expect it might be contaminated. Otherwise some contaminants may be trapped in the tank below the drain point until you level off for cruise.

Fuel Starvation - There are some design features of older planes that can cause a pilot trouble if he is not cautious.

The primitive fuel gages usually found in these planes should always be viewed with suspicion. Whether it is a cork/wire type that can easily hang up or a mechanical float type, it can lead you to believe you have more or less fuel than you actually do. Most were designed to be accurate with the aircraft in a level flying attitude so will be in error when it is in the three point attitude on the ground. The best method of checking the fuel level during the preflight inspection is to use a dipstick. Keeping track of flying time and expected fuel consumption will then give you a good check on your fuel indicator. This is especially true if you are preparing to fly an airplane you are not familiar with.

Failure to select the fuel tanks properly has caused many pilots to experience fuel starvation with fuel still in the tanks. On aircraft with more than one tank, be sure you understand the fuel system and indicators thoroughly. If your plane has one gage to read several tanks, be sure the tank being fed is also the one being read on the gage.

The design of the tank selector handle may also cause confusion. On some aircraft the handle _is_ the indicator pointing to the desired tank. On others, the pointer is across from the handle as shown in the figure on the next page.

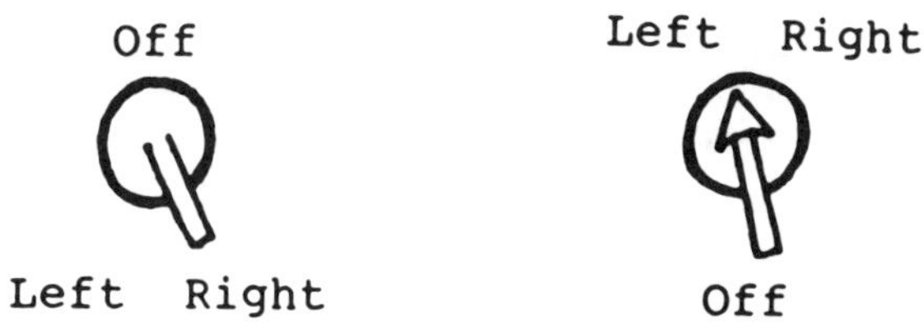

If a fuel cap has a bent tube type of vent, it should point forward. If it faces aft, fuel will be sucked out by the slipstream. Fuel caps on the wingtanks of highwinged planes are not visible from the cockpit so the loss may be undetected.

A tank vent plugged by insects, their nests, or ice can also cause trouble by creating a vacuum as the fuel level of the tank goes down. A good visual check or probing with a wire during the preflight inspection should eliminate this problem.

As a "last ditch" attempt to increase your glide after loss of an engine and a blocked fuel line is suspected, you might try slowly operating the primer pump. The priming system bypasses most of the fuel system and thus a possible blockage. Full power will not be available, but it might extend the glide to a nearby safer area.

<u>Carburetor</u> <u>Icing</u> - This is caused by the moisture in the air being cooled and freezing in the throat of the carburetor. It is cooled in two ways: by the evaporation of the gasoline and the decrease in pressure of the air passing through the venturi.

At idle, while taxiing or flying, air may be cut off to the engine by mild icing and cause the engine to lose power and possibly stop. Opening the throttle may restore power initially, but ice may continue to build. Once the fuel jets freeze over, the engine will not run again until the ice melts.

Any time the engine runs rough, suspect icing and pull the carb heat full on. Unless you have a carb temperature gage, you should use the carb heat fully on or completely off. Partial heat may raise the temperature into a range worse than the original. One indication that there was ice is a rise in RPM after the initial decrease.

Some engines, such as the Continental 65, tend to ice up more than others. Using carb heat anytime the RPM is below 2000 or if there is visible moisture in the air is often recommended.

Cruising with continuous carb heat should not damage an engine. Efficiency of the engine does go down though because warm air is less dense than cold air. Fuel consumption at cruise will increase and full power will not be available. Therfore, it should not be used if not needed.

Carb heat should be used with caution on the ground because air used with carb heat on bypasses the air filter and thus may contain dust and sand that can wear an engine badly. One exception to this rule might be its use in winter to assist in fuel evaporation while the engine warms up.

Oil Pressure Problems - During start most engines are limited to 30 seconds without oil pressure indications. If there are no indications in this time, shut down and check the oil quantity.

If the oil quantity checks OK, the oil pump may have lost its prime. This is not uncommon on older engines, especially if they have been sitting for a long time without running. Picking up the tail with the engine running (watch the propeller!) may restore the pressure. This sounds strange but I have seen it work on two different aircraft. If this fails, call an A&P mechanic for further troubleshooting. The relief valve may have hung up, a line may be plugged, or the pump may require priming with a squirt of oil through the pressure indicator line.

If the oil pressure starts to fluctuate rapidly in flight, it may be due to too low an oil level allowing the pickup tube to be uncovered momentarily. If the oil tank cap is left off or ajar, oil may be sucked out by air flow through the cowling. In any case, of course, you should land at the nearest airport and investigate. If the oil pressure goes to zero and stays there, land at the nearest suitable site with the engine off or at minimum power.

If the oil pressure steadily decreases and is accompanied by high oil temperature, there is probably a serious problem and the engine should be thoroughly checked by a mechanic before any further flight is attempted.

Ignition Problems - A proper runup check prior to flight should prevent practically all ignition difficulties by catching the problem on the ground.

Magnetoes are grounded when "off". As a result, a broken wire or bad connection will cause them to be "on". This can be dangerous during ground operation even with the engine not running as explained in the section on starting.

If this happens on an aircraft without a mixture control, it will be difficult to shut the engine off. If this happens, turn the fuel valve off and the engine will stop in a minute or two. The ignition system should then be checked while avoiding the prop arc.

If you are forced to use 100 octane gas in an engine designed for 80 octane, expect more fouled plugs than normal. See the maintenance section of this manual for troubleshooting procedures for this problem.

Fire During Start - Overpriming is the most common cause of a fire during start. If it occurs on a plane with a starter, continue cranking the engine to suck the flames back into the engine. If this is unsuccessful, turn the ignition and fuel valve off and fight the fire from outside.

For planes that must be propped, be on the lookout for signs of overpriming (such as fuel running out of the carburetor). A small fire extinguisher is a good safety item to have available in any airplane.

Fire During Flight - This is a rare occurrence, but should it happen, shut off the fuel valve and proceed with a forced landing. The fabric covering of an older type of plane can be destroyed rapidly. As you descend, try to slip the aircraft in an attempt to keep the flames away from the cockpit.

A loose oil cap may allow oil to be sucked out of the tank. If it strikes a hot exhaust stack, clouds of white smoke will be produced. The corrective procedure is the same as for any fire - land ASAP.

If your plane has an electrical system it may cause a cabin fire, often accompanied by the smell of burning rubber. Turn the master switch off and leave it off unless you need something electrical, such as the radio to land. After turning all electrical switches off (except the mags), turn the master switch back on. One by one, turn the others back on. Once you have determined which system is causing the problem, leave its switch off.

Engine Mechanical Failure - A well maintained engine will usually give considerable warning before a catastrophic failure. Abnormal oil pressure or temperature, loss of power, and unusual sounds (tapping, squealing, banging, etc) are all indications of possible impending failure and, depending on the severity, should be immediately reacted to by landing at the nearest airport or suitable site.

Another rare possibility is throttle linkage failure. If it becomes disconnected, the RPM will usually drop to idle. It then must be treated as a complete engine failure. If the linkage jams at a high power setting, set up for a forced landing and shut down the engine with the mag switch. If you have a mixture control, you may be able to set up for a more normal landing using the mixture to shut power off and on. Using the mag switch to do this will produce large backfires and may cause considerable damage.

Flight Control Problems - Fortunately this is another very rare occurrence. However, should it happen, it is not necessarily an automatic disaster. A disconnected elevator cable may require flying the aircraft with the trim system. If aileron control is lost, the rudder may be used to roll the aircraft. Most planes can be flown and landed without a rudder if there is no crosswind. Proper maintenance and preflight inspections will prevent the necessity of proving any of the foregoing statements.

Airspeed Failure - The pitot tube may become plugged by ice, a bug, or a cover may be left on (someone else must have done the preflight!). The aircraft must then be flown by feel and attitude until on the ground again. If it happens on a plane you are familiar with, you probably will have no problem landing it by feel. At a safe altitude you might review the feel of the controls when close to the stall. Land at the longest airport available. Plan on landing a little long and possibly a little fast. These planes supply many cues by feel, sight, and sound. It is not as difficult to land one of these planes without an airspeed indicator as one might expect.

<u>**Lost**</u> - If you are lost, admit it to yourself. Do not overfly an airport - no matter how small; you may not find it if you come back looking later. Land, swallow your pride, and ask where you are. If you are casual enough, maybe no one will realize you were lost!

If fuel and weather permit, climb to at least 3000 feet AGL. This increases the distance you can see prominent landmarks (cities, lakes,rivers, railroads,etc). If you have a radio this will also increase its range. Do not wander around aimlessly. Develop a plan and fly in a direction you think will bring you to an identifiable landmark. For example, here in western New York, if one flies north, he will eventually reach Lake Ontario; flying south will take him to the New York State thruway. While not the entire solution, at least it helps.

As a last resort, one can fly low enough to read road signs or names on water towers. They have saved more than one pilot from a forced landing.

If running out of fuel is imminent(say, less than 15 minutes of fuel left), select a good field for a landing. It is much better to land while you still have power and can make a low pass to check for hidden stumps, ditches, power lines, etc.

Forced Landing - If a forced landing is necessary (probably due to loss of power) keep the following points in mind :

1. Keep Control of the Aircraft. Don't get so involved in the original problem that you forget you are flying. Glide at best glide speed toward the most suitable landing site.

2. Attempt to Correct the Problem. Pull the carb heat on, check the position of the fuel valves and throttle, and cycle the mag switch.

3. Concentrate on the Pattern. Once in the landing pattern, don't lose sight of the landing field. If it is necessary to lose altitude, slip the aircraft or "S" back and forth on final approach. A 360 degree turn often loses more altitude than expected.

4. Just prior to landing, switch off the mags to prevent a burst of power causing you to overshoot the landing site. Prepare for a soft field landing and "straightarm" the instrument panel to keep from going forward into the controls. The resulting forward stick would raise the tail and cause the airplane to flip onto its back.

One of the most important elements of a successful forced landing is the selection of a landing field. Pastures, recently cultivated fields with low crops, athletic fields (such as golf courses), and roads are all good possibilities. Fields to avoid if possible: ones with high crops such as corn or wheat which can hide hazards or cause groundloops, plowed fields unless you can land parallel to the furrows, and ones with standing water.

If landing on or near roads, watch for signs and powerlines. Four lane highways and the area between lanes often are free of powerlines and are wide enough for many planes. It is usually illegal to land on roads, but if it is the only place to go in an emergency, who cares?

Some other factors that may affect your selection of landing sites are winds, the slope of the land, and your altitude. Obviously, the lower you are the shorter your glide range will be. You should keep the possibility of a forced landing continually in mind as you fly. If over rough terrain, you should fly higher than if over farmland.

Wind is important in two ways - the effect it has on your glide range and the desirablity of landing with a headwind. Some common indicators of wind direction and strength are smoke, flags or clothes,and ripples in water or crops.

If there is a definite slope to the land, it is best to land uphill even if it means landing with a some tailwind.

Once you decide on a landing site, don't change your mind unless absolutely necessary. Stretching the glide may result in a stall or not reaching the intended spot.

The pattern you fly prior to a forced landing varies with wind and your initial altitude. If you have sufficient altitude, you should fly a downwind, a base, and a final as in a normal landing. Practice will teach you how to vary the pattern to account for different circumstances. The following diagram shows how the standard pattern can be varied to place the aircraft properly to touch down at the desired point :

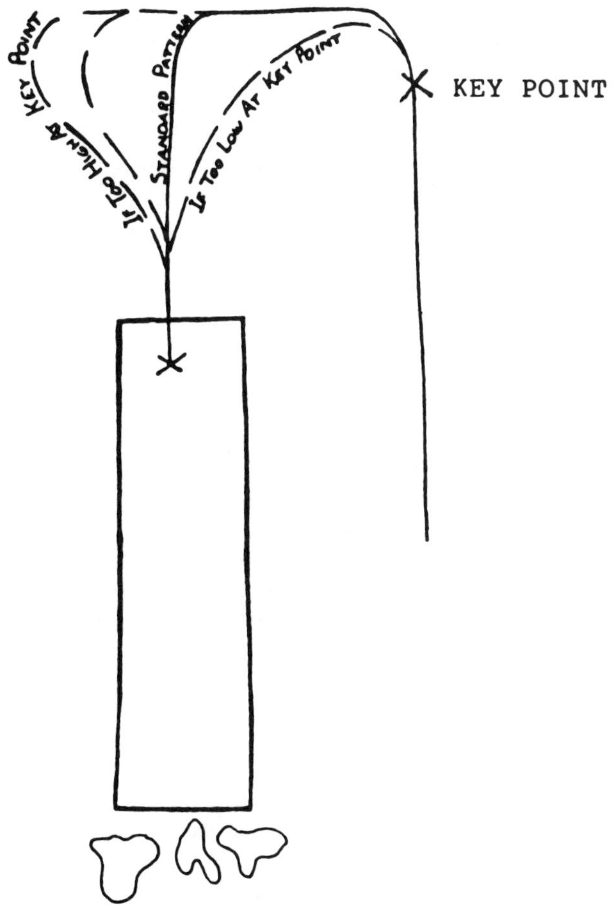

REGULATIONS

This section is a short summary of FAA regs that apply to taildraggers. Most apply to all planes but were included because of the type of flying these basic planes often do. Some are simplified to help you remember them. It is by no means a complete list.

Minimum Weather - There are many rules governing VFR weather, but if you avoid towns, 1000 foot ceiling and three mile visibility will keep you legal almost anywhere. That is also a practical, safe minimum for flying these basic planes.

Minimum Altitudes -

a. over towns or cities - 1000 feet above obstacles within 2000 ft.

b. over other areas - except for landing, stay at least 500 ft AGL. Over water or sparsely populated areas, stay 500 ft from people, vehicles, and buildings (no altitude restriction).

No Radio - Stay away from TCA's and ARSA's. Boundaries and altitudes are shown on Sectional charts. See a flight instructor if you aren't familiar with them. All other airports with a tower must be avoided by five miles unless you are above 3000 ft AGL or are landing at another uncont-rolled airport within that five mile circle.

Right of Way - Anything with less maneuverability than you has the right of way. Examples are gliders or balloons. If two airplanes are converging at the same altitude, the one on the right or the one being overtaken has the right of way.

Aerobatics - Limited aerobatics are legal in some of these planes because they were certified under different rules than planes built today. However,many of them are over forty years old. Unless you are certain of the structural soundness and are competent in the maneuvers, aerobatics aren't recommended.

Night Flying - A plane can not be legally flown from sunset to sunrise unless it is equipped with an anticollision beacon and navigation lights.

To Carry Passengers - You must have made three takeoffs and landings in the past 90 days. In a tailwheel type plane, they must have been to a full stop - not touch and goes.

Don't Forget - You need a Biennial Flight Review (BFR) endorsement by a flight instructor in your logbook. It must have been accomplished within the past 24 months.

Third Class Medical Certificates are only valid for two years.

TECHNICAL TOPICS

Stalls and spins - The single reason a wing stalls is the angle of attack (the angle between the wing and the airflow) is too great. Although this may occur because the airspeed is too low for the amount of lift you are asking for, it can happen at any speed. Angle of attack is controlled by the elevator. Therefore, the most important step in recovering from a stall is to get the control wheel/stick forward. Full power to increase airspeed will also help.

Use of ailerons while these planes are stalled must be minimized. If a wing drops, the common reaction is to apply aileron opposite the bank. Lowering the aileron on the low wing increases the angle of attack on that wing insuring that it will remain stalled. The aileron on the high wing will be up, decreasing its angle of attack, probably below stall. As a result the low wing will go lower. Also, the drag of the stalled wing will be greater, resulting in a yaw and possible spin. If this happens near the ground (often combined with an attempt to hurry the turn to final with rudder) a stall/spin accident results. The controls should be placed forward and excess rudder should be removed.

Turns - Most of these planes have more adverse yaw (the tendency to yaw away from a lowered aileron) than more modern planes do. This makes them better trainers of the proper use of the rudder. An interesting demonstration can be performed in these planes by applying full roll control and releasing it while keeping the rudder pedals centered. The plane will yaw away from the bank an amount equal to the turn due to the bank and will end up close to the original heading. For pilots trained on modern trainers this can be a real "eye opener" about adverse yaw.

Another strong effect on these planes is the tendency to roll away from airflow from the side. This is called dihedral effect and causes a right roll when the right rudder is pushed and left roll with left rudder. The plane first yaws in the direction of the depressed rudder pedal, causing the airflow to strike the plane from the opposite side. This in turn causes a roll in the direction of the depressed rudder pedal. One can thus control the bank angle of the plane using only rudder if necessary.

There is a tendency for the bank to increase in a steep turn. It may require aileron toward the outside of the turn to keep from developing an increasingly steep bank. This is due to the outside wing traveling in a larger circle than the inside one in the same time resulting in a faster speed. This develops more lift and results in a rolling moment toward the inside of the turn.

Left Turn Tendency - With high power settings and at speeds less than cruise, there is a tendency for the plane to turn left. Sometimes erroneously labelled "torque", it is caused by "P" factor and the corkscrew path of the airflow from the propeller striking the left side of the tail.

"P" factor is the moment caused by the increased thrust produced by the right or downgoing propblade (an airfoil shape) and the decreased thrust on the left or upgoing blade.

These combined forces must be counteracted by right rudder or a constant lefthand turn will result. On an aircraft with a propeller that turns counter-clockwise the forces are in the opposite direction and the plane will tend to turn right.

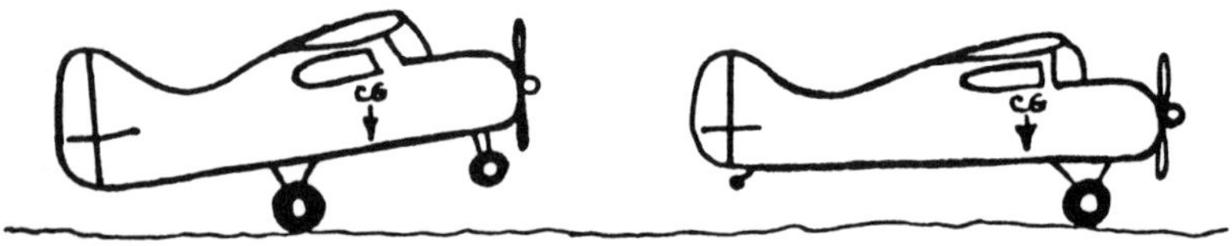

Forces During Landing - Any descent rate at touchdown with a tricycle geared plane will cause the nose to pitch down, decreasing its angle of attack, and keeping it on the ground. The same descent rate with a tail-wheel plane will cause the tail to drop, increasing the angle of attack, and probably result in a bounce. This can be minimized by landing at the proper speed so all wheels touch at the same time or tailwheel first.

Now let's look at the two types of aircraft from above :

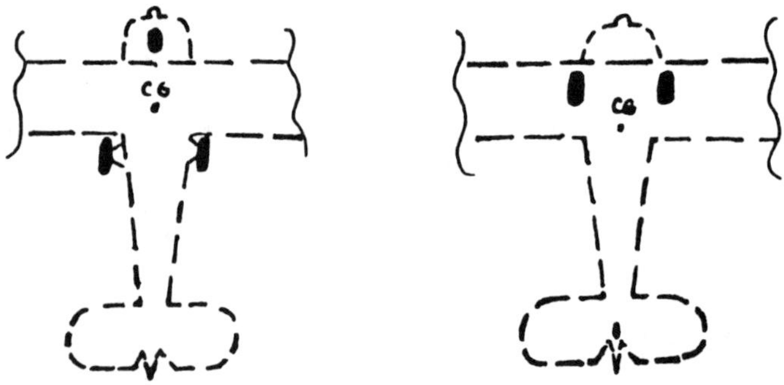

If each lands on its main gear while in a drift, the nosegear plane will tend to align with its direction of travel. The tailwheel plane, however, tends to turn away from the line of flight. It may not be too strong a tendency for proper use of rudder to correct, but it sets up a possibility of over-correcting. Practicing landings on a grass runway is a good idea because a tire will slide slightly on the grass before transmitting the sideforce to the plane.

Keeping a tailwheel plane going straight on the ground is like balancing a stick on your finger. The further from the center it gets, the more it tends to keep going. Small corrections early are much better than large corrections later. If a tailwheel plane has a drift so large that the CG is outside the track of the mainwheels, even differential braking will not prevent a ground loop.

"Rules of Thumb" -

a. You can estimate the approximate fuel consumption of an engine (at 75% power) by multiplying its horsepower by .06 .

b. Other things remaining equal, it takes a large increase in horsepower for each increase in airspeed. For example, to double a plane's speed solely by an increase in engine horsepower will require eight times the power! Reducing the drag of an airframe is a more efficient means of higher speed than installing a larger engine. A larger engine will result in a better climb and shorter takeoff distance though.

c. The best speed to fly for max range (no wind) is approximately the same as the best glide speed.

NUMBERS TO KNOW

This section lists performance figures and limitations of nine of the most common classic taildraggers. The numbers were obtained from original owner's manuals, personal experience, and estimated when necessary from certain relationships that hold for nearly all planes. Other factory planes and homebuilt aircraft of this general type can be estimated from these examples.

The weights given are approximate. Check the aircraft weight and balance sheets for your specific aircraft. Airspeeds given are calibrated so may differ slightly from what you see on the gauge due to system errors. Speeds are also determined by the type of propeller installed. An aircraft used for training is apt to have a climb propeller so will not cruise as fast as one equipped with a cruise prop.

All specs are meant as a general guide for use when no other information, such as from a pilot familiar with the plane, is available.

AERONCA CHAMP

This fabric covered, tandem seat aircraft is controlled with a stick. Its fuel tank is located ahead of the instrument panel. It has an oleo strut type landing gear, a standard door on the right side, and a trim tab on the elevator. Early models had 65 HP engines, but were often modified to use 75 or 85 HP. Later models had the larger engines. They were the forerunner of the Citabria. Two distinguishing features of a Champ are its round vertical tail and bubble windscreen. 85 HP models require a larger vertical stabilizer. It is soloed from the front seat. The following numbers apply to the 65 HP models.

LOADING:	Gross Wt	1220
	Useful	510
	Fuel	78
	Oil	8
	Baggage (solo)	40
	(dual)	20

OIL:	4 qts Summer SAE 50	
	Winter SAE 30	
	Normal Pressure	30 - 40 psi
	Max Temperature	220 F

FUEL:	Quantity	13 gals
	Consumption	4 - 4.5 gph
	Endurance	2.5 hrs
	Range	190 mi

RPM:	Idle	550-600	Static	2010-2200
	Runup	1800 (75 max drop)		
	Cruise	2150	Never Exceed	2300
	Shutdown	800		

SPEEDS:	Liftoff	50-55	Best Angle Climb	55
	Best Rate Climb	60	Cruise	75
	Power Off Stall	45	Final Appch	60
	Never Exceed	129	Short Field	55

CITABRIA

Citabrias have a strong family resemblance to the Champ. Differences include a squared-off tail, longer engine cowl, prop spinner, spring type landing gear,larger side windows, and fuel in the wings. The 7 ECA has 115 HP, the 7 GCBC has 150 HP and flaps, and the 7KCAB has 150 HP and no flaps. All are aerobatic. Those with fuel injected engines have inverted flight capability. Much of the useful load is taken up by fuel so flight with a passenger often must be done with less than full tanks. The following numbers are for the 150 HP models.

LOADING:	Gross Wt	1650	
	Useful	510	
	Full Fuel	210	
	Oil	15	
	Baggage	100	(normal category)

OIL:	8 qts	Above 60 F -	SAE	40-60
		0 - 60 F	SAE	40
		Below 0 F	SAE	30

FUEL:	Quantity	35	gals
	Consumption	8.5	gph
	Endurance	4	hrs
	Range	480	mi (no reserve)

RPM:
Warm up 1000-1200 Static 2200+
Runup 1800 (200 max drop)
Cruise 2600 Never Exceed 2700

SPEEDS:
Liftoff 55-60 Best Angle Climb 58
Best Rate Climb 69 Cruise 120 TAS
Power off Stall (flaps up/dn) 51/45
Final Appch 60-70 Best Glide 65
Never Exceed 129 Short Field 55
Max Flap Speed (GCBC) 90

CESSNA 120 & 140

These two aircraft are quite similar. The 120 has a metal fuselage, fabric wings,and an 85 HP engine. Later model 140's are all metal, flap equipped, and some have 90 HP. Identification features include spring gear, round wingtips, wheel control, and horizontal bars across the engine cooling air intake openings. Early models had "vee" lift struts but the later ones had a single strut. Often the early models were modified by adding gear "extenders" to improve ground handling. They move the main wheels four inches ahead of the original position. Spins and slips with flaps down are prohibited. Use of carb heat until takeoff may be necessary.

LOADING:	Gross Wt	1450
	Useful	650
	Full Fuel	150
	Oil	8
	Baggage	80

OIL:	4.5 qts	Above 70 F	SAE 40
		32 - 70 F	SAE 30
		Below 32 F	SAE 20

Normal Pressure 30-40psi (10 @ idle)
Max Temperature 220 F

FUEL:	Quantity	25 gals (1/4 tank min for TO)
	Consumption	5 gph
	Endurance	5 hrs (no reserve)
	Range	450 mi " "

RPM: Idle 500-550 Static 2100
Runup 1800 (75 max drop) Cruise 2150
Never Exceed 2300

SPEEDS: Liftoff 50 Best Angle Climb 60
Best Rate Climb 75 Cruise 90
Power Off Stall (flaps up/dn) 49/45
Final Appch 65-70 Short Field 60
Never Exceed 140 Max Flap Speed 82

CESSNA 170

This plane looks much like an early 172 except it has a tailwheel and a rounded vertical tail. Early models had fabric wings with rounded tips. Later models had all metal wings with tapered, squared off tips. All have spring landing gear and flaps. The standard engine is a 145 HP Continental. Spins and slips with full flaps are prohibited. Use of carb heat may be necessary until takeoff.

LOADING:	Gross Wt	2200
	Useful	980
	Full Fuel	222
	Oil	15
	Baggage	120

OIL: 8 qts Above 50 F SAE 40
Below 50 F SAE 20
Normal Pressure 30-40psi (5 @ idle)
Max Temperature Red Line

FUEL:	Quantity	37 gals
	Consumption	8 gph
	Endurance	4.5 hrs (no reserve)
	Range	540 mi " "

Do not takeoff with needle in red area on gauge of selected tank.

RPM: Idle 600 Static 2230-2330
Runup 1600 (100 max drop)
Cruise 2200-2450 Never Exceed 2700
Shutdown 800

SPEEDS: Liftoff 60 Best Angle Climb 75
Best Rate Climb 85 Cruise 120
Power Off Stall (flaps up/dn) 58/52
Final Appch 70-75 Max Flap Speed 100
Short Field 60 (full flaps)
Never Exceed 160

LUSCOMBE

Early models (8A) have 65 HP engines and fabric wings. Later models have 65 or 85 HP and fabric or metal wings. Many have electrical systems. All are stick controlled, side-by-side planes with a door on both sides. They look much like the Cessna 120 or 140. Two features that can help identify them are the landing gear (Cessna has a spring while the luscombe has a "vee" strut) and engine cowlings (the Luscombe has a "chin"). Often brakes were installed only on the left side. Limited aerobatics are authorised after installation of a fuselage stiffener. The following numbers apply primarily to the 8E model.

LOADING:	Gross Wt	1400		
	Useful	550		
	Full Fuel	180		
	Oil	8		
	Baggage	75		
OIL:	4.5 qts	Above 70 F	SAE 40	
		32 - 70 F	SAE 30	
		Below 32 F	SAE 20	
	Normal Pressure		25 - 48 psi	
	Max Temperature		220 F	
FUEL:	Quantity	30 gals		
	Consumption	5.5 gph		
	Endurance	5 hrs		
	Range	500 mi		
RPM:	Idle	550	Static	Approx 2100
	Runup	1700 (50 max drop)		
	Cruise	112	Never Exceed	145
SPEEDS:	Liftoff	50-55	Best AngleClimb	60
	Best Rate Climb	72	Cruise	100
	Power Off Stall	48	Final Appch	70
	Never Exceed	145	Short Field	60

PIPER CUB (J3)

This is THE classic taildragger and its price reflects it. Its distinguishing features are: tandem seats, bungee cord landing gear, a two flap door on the right side, and aileron cables running up the lift struts. Pitch trim is supplied by raising or lowering the leading edge of the horizontal stabilizer. The original engines were 65 HP, but many have been modified with larger ones. The cylinders project out of the cowling into the airflow for better cooling.

The PA-11 is nearly the same except it has a fully enclosed engine and fuel (18 gal) in the wing instead of 12 gal ahead of the instrument panel. Most J3's are soloed from the rear. The PA11 is soloed from the front.

LOADING:	Gross Wt	1220
	Useful	540
	Full Fuel	72
	Oil	8

OIL: 4 qts 70 - 100 F SAE 40
32 - 70 F SAE 30
Below 32 F SAE 20
Min Pressure 30 psi/Max Temp 220 F

FUEL:	Quantity	12 gals
	Consumption	4 - 4.5 gph
	Endurance	2.5 hrs
	Range	190 mi

RPM: Idle 550-600 Static 2070-2250
Runup 1800 (75 max drop)
Cruise 2150 Never Exceed 2300
Shutdown 800

SPEEDS: Liftoff 50-55 Best Angle Climb 50
Best Rate Climb 55 Cruise 75
Power Off Stall 38 Final Appch 55
Never Exceed 122 Short Field 50

PIPER PACER AND MODIFIED TRI-PACER

These planes are nearly identical except, originally the Tri-pacer had a nosegear instead of a tailwheel. There is a popular STC available to convert them back to tail-draggers. They are wheel controlled and seat four people. They have shorter wings than Cubs and most are equipped with flaps. Early models had 125 or135 HP, but later ones had 150 or 160HP.

LOADING:	Gross Wt	2000	
	Useful	940	
	Full Fuel	216 (std)	264 (opt)
	Oil	15	
	Baggage	50	

OIL: 8 qts Above 40 F SAE 50
Below 40 F SAE 40
Normal Pressure 65-85 psi
Max Temperature 245 F

FUEL: Quantity 36 gal (std) 44 (opt)
Consumption 9 gph
Endurance 4 - 5 hrs no reserve
Range 500 - 625 mi " "

RPM: Idle 550-600 Static 2270-2480
Runup 1800 (100 max drop)
Cruise 2450 Never Exceed 2700
Shutdown 800

SPEEDS: Liftoff 55-60 Best Angle Climb 70
Best Rate Climb 84 Cruise 125
Power Off Stall 49 (flaps dn)
Final Appch 75-80 Max Flaps 95
Never Exceed 145 Short Fld 65

PIPER SUPER CUB (PA-18)

This model has many of the external features of the J-3 except it has a fully cowled engine, "balanced" elevators, and most have flaps. It has fuel in the wings (36 gals) and is soloed from the front. Early Super Cubs were lighter, had no flaps, and smaller engines. Engine sizes were 95, 115,125,135,150, and there is an STC now for 180 HP! There have been so many modifications for it, several of the following numbers may not apply to your specific plane.

LOADING:	Gross Wt	1750		
	Useful	820		
	Full Fuel	**216**		
	Oil	15		
	Baggage	50		
OIL:	8 qts	Above 60 F	SAE 50	
		30 - 90 F	SAE 40	
		0 - 70 F	SAE 30	
		Below 10 F	SAE 20	
	Normal Pressure	65-85 psi		
	Max Temperature	245 F		
FUEL:	Quantity	36 gals		
	Consumption	9 gph		
	Endurance	4 hrs		
	Range	460 mi		
RPM:	Idle	550-600	Static	2350-2450
	Runup	1800 (75 max drop)		
	Cruise	2400	Never Exceed	2700
	Shutdown	1200		
SPEEDS:	Liftoff	50	Best Angle Climb	55
	Best Rate Climb	75	Cruise	115
	Power Off Stall	45	Final Appch	65
	Never Exceed	153	Short Field	55

STINSON

The "Voyager" model of this four place taildragger can carry 350 lbs of baggage with the rear seat removed. The "Station Wagon" model has a wood and vinyl interior and can carry 600 lbs in the rear! Both have rudder trim and slots in the leading edge of the wing. More elevator travel is allowed when the flaps are down than when they are up. Early models had 150 HP. The later ones had 165 HP. The button above the mag/starter switch must be pushed to allow the switch to be placed in the "start" position. Carb heat knob must be rotated before it can be pulled. The following numbers are for the 165 HP models.

LOADING: Gross Wt 2400
Useful 1100
Full Fuel 275
Oil 17

OIL: 9 qts Above 40 F SAE 40
Below 40 F SAE 20
Normal Pressure 35-55
Max Temperature - Red Line

FUEL: Quantity 46 gals
Consumption 10 gph
Endurance 4.5 hrs
Range 500 mi (no reserve)

RPM: Warm up 1000 Static Varies
Runup 2000 (200 max drop)
Cruise 2450 Never Exceed 2800
Shutdown 800

SPEEDS: Liftoff 65 Best Angle Climb 70
Best Rate Climb 80 Cruise 110
Power Off Stall (flaps up/dn) 65/62
Final Appch 80 Short Field 70
Never Exceed (norm/utility) 158/170
Max Flap Speed 88

TAYLORCRAFT

This plane with side-by-side seating, wheel controls, and a bungee landing gear suspension carries its fuel ahead of the instrument panel and in a six gallon tank in the wing. The fuselage is noticably narrower just ahead of the tail than those of its contemporaries. They are efficient cruising planes, but don't command the prices that Cubs do. They have a reputation for floating during landing. This might be expected if the pilot uses the owner's manual final approach speed - 25 mph over stall speed!

LOADING:

Gross Wt	1200
Useful	500
Full Fuel	108
Oil	8
Baggage	50

OIL:

4 qts Warm Weather SAE 40
Cold Weather SAE 30
Normal Pressure 10 - 35 psi
Max Temperature 220 F

FUEL:

Quantity18 gals(12 in main, 6 in wing)
Consumption 4-4.5 gph
Endurance 4 hrs
Range 350 mi

RPM:

Idle 550-600 Static 2070-2250
Runup 1800 (75 max drop)
Cruise 2150 Never Exceed 2300
Shutdown 800

SPEEDS:

Liftoff 45 Best Angle Climb 50
Best Rate Climb 55 Cruise 85
Power Off Stall 35 Final Appch 60
Never Exceed 140
Short Field 45 - 50

MAINTENANCE

There are several maintenance operations the FAA has authorized the owner/pilot to perform without being an A&P mechanic or being supervised by one. He must hold at least a Private Pilot Certificate and enter a description of all work he performs in the engine/airframe logbook. After the entry, he must sign his name and include his pilot certificate number. It probably would be a good idea if he were supervised the first time he does each one as damage can be done while performing the simplest task.

A list of authorized preventative maintenance steps follows :

1. Removal, installation, and repair of landing gear tires.
2. Replacing elastic shock absorber cords on landing gear.
3. Servicing shock struts with air or oil.
4. Cleaning and greasing wheel bearings.
5. Replacing defective safety wiring or cotter keys.
6. Lubrication not requiring disassembly.
7. Making simple fabric patches (no rib-stitching or structural removal).
8. Replenishing hydraulic fluid.

9. Painting (excluding balanced control surfaces).
10. Applying preservatives to components.
11. Repairing upholstery without disassembly of primary structure.
12. Repairing fairings, cowlings,other nonstructural covers.
13. Replacing side windows.
14. Replacing seatbelts.
15. Replacing seats or seat parts without disassembly of primary structures.
16. Repairing landing light circuits, bulbs, lenses, and reflectors.
17. Replacing wheels and skis with no weight and balance required.
18. Replacing cowling not requiring removal of prop or flight controls.
19. Replacing, cleaning, regapping spark plugs.
20. Replacing any hose connection except hydraulic connections.
21. Replacing prefabricated fuel lines.
22. Cleaning fuel and oil strainers.
23. Checking and replacing batteries.

If a tire wears on one side more than the other for any reason, it should be reversed on its wheel before any fabric shows through. CAUTION - make certain the tire is completely deflated before loosening the bolts holding the wheel halves together! A rule of thumb for tire pressure is approximately 1 psi for each 100 lbs of gross weight.

When moving the plane on the ground, don't let anyone push on the tail stabilizer rods, prop, or the middle of the wing lift struts. Use the lift handle to pick up the tail, not the stabilizer. To pick up one main gear to check for a dragging brake or to place a block under it to change a tire, pull down on the opposite wing where the struts end - NOT the wingtip bow. At the same time someone else should push up on the wing to be raised. This will eliminate the need for a jack on a small plane.

Another procedure the owner can do is clean, gap, and rotate spark plugs. Replacing the plugs from the bottom of cylinders with those from the top, and vice versa, may help reduce plug fouling. Cylinder heads are aluminum so be very careful not to overtorque plugs when replacing them. Use a torque wrench and follow the manufacturer's recommended limit. Use a SMALL amount of antiseize compound on the threads to ease plug removal next time.

If you discover an excessive RPM drop and roughness during the pretakeoff runup check, suspect carburetor icing first and apply carb heat. If that does not cure the problem, probably a spark plug has become fouled. If your plane has a mixture control, lean the engine and run it at a high power setting for a short time to clear the fouling. If you do not have a mixture control or if this technique does not cure the problem, determine which plug is fouled by running the engine on the mag producing the roughness. After a minute or two, check for a cold cylinder with a wet rag or CAREFULLY with your hand. By checking which plug (top or bottom) is fired by the selected magneto, you can identify the fouled plug without a long trial and error process.

The air cleaner is a very important accessory on an aircraft, but is often neglected. Dirt particles that pass through a wornout air cleaner can drastically shorten engine life. On old style air cleaners, if the metal screen is visible, it is time to change it. If you can see light through it, a change is past due.

If your plane has a wooden prop, it is best to leave it in a horizontal position and insure that the drain holes in the tips are clear. This will prevent moisture from collecting in one blade and making the prop out of balance. Also, if the air is appreciably drier or more moist than when the mounting bolts were last torqued, they should be checked again. If it is drier, the wood may shrink requiring the bolts be tightened. If it is more moist, the wood may swell and crush some of the fibers under the hub.

If you have a metal prop, there is not much to do except occasionally rub a light coat of oil on it and have an A&P file any nicks out of it as soon as you find them.

Fabric work is expensive because of the large amount of labor involved. It is not difficult to learn and can best be done by working with an experienced fabric person. Often the most work required is the removal of corrosion, repair of any damage, and priming of the underlying structure.

If you are planning on repainting your plane, without recovering it, remember that enamel can be applied over dope, but dope can not be applied over enamel. Most dark colors will "bleed through" lighter colors applied over them. Red dope is the most apt to do this and it is the most expensive color to buy. Dark colors will absorb more ultraviolet rays than light colors so fabric will deteriorate faster under dark colors. If cotton tape is applied over Ceconite fabric, it often deteriorates faster than the base fabric. If your tapes appear shoddy, you might consider replacing only the tapes.

Anytime you punch test fabric be sure to remove all paint at the test spot - especially if it is enamel. Otherwise you are apt to get an artificially high reading.

The interior of wings should be checked occasionally for wasp nests and mouse damage. Wasp nests can break loose and jam aileron pulleys. Mice will chew the ribstitching threads to allow themselves free movement and sometimes use them to make nests. If you discover this, you will have to restitch the damaged area. Mouse urine is also very corrosive. If you don't mind the smell, a few mothballs out in the wings will keep mice away if you are going to leave the plane idle for a while.

To remove an inspection cover on a fabric plane, press in the center until it "pops", slide it parallel to the clip on the inside until one end clears the side of the hole, and lift it out. When you replace it, there is less chance of losing it in flight if you position the retaining clip perpendicular to the airflow.

PK screws will stay tight longer if they are shellacked before being replaced. After a time, the next larger size will be needed to hold fairings.

Clean plexiglass only with water or special plastic cleaner. A clear wax may fill in small scratches, but test it on a small, inconspicuous area to make sure it is compatible with the plastic and doesn't cause crazing or fogging. There are kits available to allow you to sand out fairly large scratches. If a crack appears in a window, drill a hole in the very end of it. This will relieve the stresses that caused it and keep it from continuing. A drop of acetone applied with a toothpick will help seal it.

A lot of corrosion can be avoided by properly draining the airframe. Be sure the drain holes in all surfaces are clear and in the right places. If in doubt put another one in. Be particularly careful that the area around the tailpost of the fuselage drains well. This is usually the lowest point of the fuselage of a taildragger. Corrosion is very expensive to repair in this area due to the amount of welding required to replace tubing.

Of course, flight controls deserve close attention. If you follow your owner's manual and lubricate flight controls with oil, use as little as possible. Excess oil will trap dirt and cause faster wear. Powdered graphite will solve that problem, but leaves black streaks across the control surfaces. To check for wear at elevator hinges, place the elevator up into flying position and push the leading edge of it up and down. If you leave it down in its normal resting position you won't see the most looseness. Often wear in a control can be cured by replacing the pin instead of the hinge as the pin should wear faster. If that is unsuccessful, the hinge must be replaced or a new bushing installed. Unfortunately, not all planes use bushings.

If there is excessive play in the sticks in a tandem seated plane, the one not used for solo flight usually wears faster. The pilot's hand acts as a shock absorber and doesn't allow as much vibration. Excessive wear of the linkages can be cured by new bolts, bushings, or complete reworking. In any case it is a job for an A&P mechanic.

Many of the planes we are discussing are forty years old and have accumulated several AD notes (Airwothiness Directives). Compliance with these mandatory inspections or modifications should be correctly documented in the maintenance log books and Form 337's, if required. Check that they have been done and not just signed off. Many A&P's and all Authorized Inspectors can show you a list of those that affect your plane.

Parts for the more common older planes are still available, if not locally then through "Trade A Plane". Univair and WagAero can supply parts for most classic taildraggers, too. Addresses can be found at the end of the "Buying" section of this manual.

Three handy items that you can build easily are gust locks, bird guards, and a tiedown rope to restrain your plane when you start it by hand propping.

Gust locks keep flight control surfaces from blowing in the wind when the aircraft is tied down outside. You can imagine the wear on the the hinges during weeks of oscillating in the wind if they are left free. Using the seatbelt as described earlier to hold the elevator is a good idea, but doesn't hold the ailerons very well and does nothing for the rudder.

Excellent gust locks can be made by padding wooden squares and connecting them with a bolt and a wing nut as shown in the figure on the following page. If the bolt is small enough, slide it between an aileron and the wing. Of course, you only need one. Put a LARGE red streamer on it so you won't fail to remove it before flight!

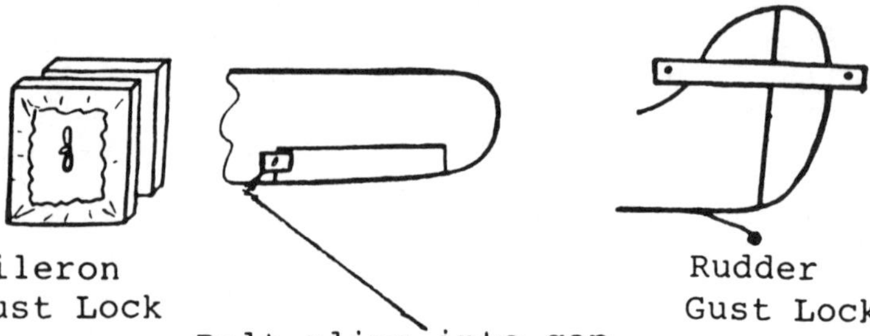

Aileron Gust Lock

Bolt slips into gap between aileron & wing

Rudder Gust Lock

The rudder gust lock is made from two strips of wood, padded if you wish, long enough to slip over the fin and rudder as shown and connected loosely with bolts. Again, use a red streamer to remind you to remove it before flight.

Birds love to build nests in these older planes. There are usually a lot of openings for them to enter. Especially in the spring, it is worthwhile stuffing foam rubber into openings where birds could possibly get in. Incidentally, make sure there isn't one in there when you close the opening.

Some planes can benefit from a bird guard for the cowling made by gluing styrofoam or foam rubber to a "Y" made of plywood as shown. The red "flag" is to remind you to remove it before starting the engine.

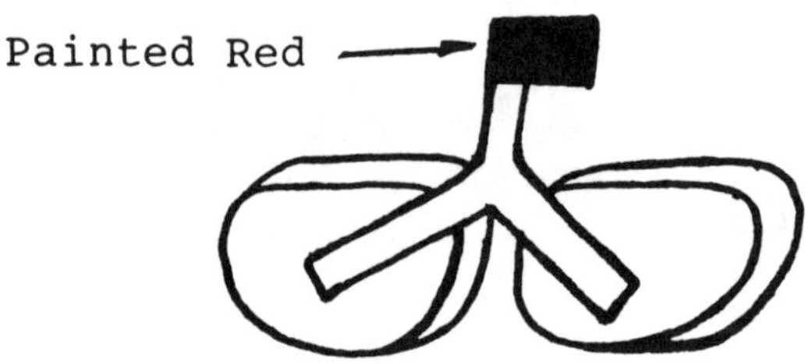

As emphasized in the section on starting, you should have a knowledgable person in the cockpit when handpropping an aircraft. Unfortunately, there may be a time when you are by yourself or no one with you knows anything about planes. Glider tow hitches can be used to hold a plane back until the pilot gets in his seat after propping the plane. He can then release the hitch and taxi away.

An alternative to a glider hitch is a rope with a loop in each end as shown in this drawing:

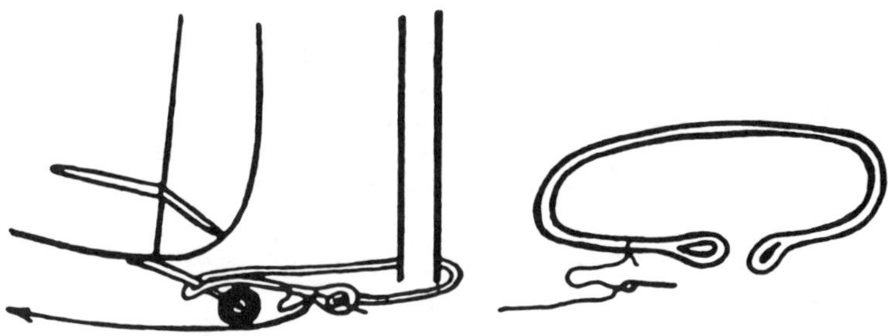

To Cockpit

It is passed around a sturdy object such as a pole and the tailwheel leaf spring. One loop is pushed part way through the other loop and a metal pin is used to secure them together. After the engine is started, the pilot climbs in and pulls the pin by means of a string attached to it. If the string is also tied to the rope, he can retrieve it without leaving the cockpit. This is only a stopgap method and should only be used if no other means of tiedown is available.

Inactivity is very hard on an aircraft. The best procedure to keep all systems in good condition is an occasional short flight. This will heat the engine to operating temperature, evaporate moisture, and supply complete oiling.

Do not just run the engine on the ground! More harm than good is done by this occasional ground running. The cylinders can overheat at high RPM with little cooling airflow and the oil won't get warm enough to evaporate the moisture and gas accumulated in the sump. On the other hand, a short flight exercises all systems without the disadvantages.

If you plan on storing your plane, even if just for the winter, there are steps you can take to help prevent damage.

Remove as much dirt as possible from the interior of the plane. This minimizes corrosion caused by the moisture trapped in the dirt. A good coat of wax will help protect the exterior finish and will make it easier to clean after storage. An automotive upholstery protectant, such as Armor All, will keep your cushions pliable and looking good. Plug any openings in the plane where rodents might enter with rags or foam rubber. Moving the plane occasionally to prevent "flat spots" on the tires from developing is a good idea if you don't want to go to the trouble of blocking it up with the weight off the wheels. If the plane has a battery, remove it and store it in a warm place.

If you are storing the plane longer than a few months, see an A&P mechanic for special steps to oil cylinder walls, purge the fuel system, and replace the spark plugs with special dehydrating plugs,etc.

BUYING A TAILDRAGGER

There are many books available on the topic of purchasing an airplane. Although they are directed more toward buying fairly recent types such as Cessna 172's, Piper Cherokees, etc, they contain much useful information for the prospective taildragger buyer, too. Therefore, this chapter will describe items unique to the purchase of older aircraft.

Although the majority of sellers will be honest, there are a few unscrupulous individuals who might alter paperwork, cover up shoddy maintenance, or pull other "dirty tricks". Your best protection is to be alert and knowledgeable.

Educating Yourself About The Plane

Before beginning the search for a specific plane, obtain as much information as possible about the model you are interested in. A good place to start is by purchasing an owner's manual for it. Wag-Aero and Univair are two sources for manuals for the older planes. Their addresses are listed at the end of this section. The owner's manual will show some of the design and construction details. Be skeptical about their published performance figures though - they were notoriously optimistic. Some gave maximum range figure based on flying until the engine quit from lack of fuel rather than the modern practice of allowing for a small reserve.

Owners can give you more realistic figures as well as other advice about cost of operation, maintenance, supply of parts, etc. They may also know of an owner's club for the type of plane you are considering. These clubs usually can supply buying hints and most have lists of planes for sale. The information may be a little biased, but is still useful.

Another excellent source of information on older aircraft is the "old timer" mechanic who has worked on them for years - maybe since they were new! They are getting scarce, but can sometimes be found at the small, out of the way airports around the country. It is well worth the small fee they may charge to inspect any plane you are seriously considering buying.

To determine the price range of any type of plane, check Trade A Plane and local pilots and mechanics. Keep in mind advertised prices are "asking prices and actual selling prices may be considerably lower. Commonly, ten per cent is added to what the seller really wants in order to have "something to bargain with". Offering about 20% less than the advertised price will allow you to raise your bid later until you arrive at a mutual agreement. Not all sellers play this game, but you never know which ones do until you try.

When discussing prices with local pilots or mechanics be sure you are talking about recent prices as they tend to increase each year. You can plan also on hearing hangar stories of the "creampuff" Cub that was just sold for $2500. Don't believe it.

The "blue book" price guides are often only accurate for newer planes and show absurdly low prices for older, classic ones. Most don't even list them.

The largest determining factor of price on older planes is condition - particularly of the fabric and engine. Few have any avionics, but if one does you can see what a $1000 radio does to the price of an airplane that would otherwise cost $5000. By the way, radios can be a good bargaining tool. If the seller is trying to impress you with how valuable his radio is, offer to buy the plane without the radio for a price reduced by what he thinks the radio is worth. Usually he won't do it, but it does put the whole thing in perspective.

Locating The Plane

The first place to look for a specific plane to buy is locally - newspapers, airports, and tips from friends. Shopping locally has two advantages. First, you can inspect the aircraft at little cost without committing yourself to a long trip or paying a deposit to have it flown to you. If it is necessary to travel far to inspect an aircraft, there is a temptation to accept a less than optimum deal just because you don't want to "waste the trip". More about this later.

The second advantage of buying locally is that the best deal can often be found close to home. You are able to "wait out" an offer, you don't have inspection trip and ferry expenses, and you can have an A&P mechanic that you know inspect the plane. If an exceptionally good deal is offerred locally it will usually be sold before it is nationally advertised. Also, once the word gets around that you are in the market for a plane, you may be approached by an owner who hasn't even tried to sell his plane before.

Nationwide sources have the advantage of a larger selection. If you have your heart set on, say, a 1947 blue Funk you will have to search the entire country for it. There is the old standby, Trade A Plane, plus several new advertising papers and magazines that have at least some older planes advertised. Don't be frightened away from long distance dealing by the possible difficulties. They can be minimized with a little forethought.

If the distance isn't too great, offer to pay the owner's expenses to fly the plane to your airport for inspection. If you buy the plane he may not even charge you these costs.

If you must travel very far to see a plane, explain the problem to the seller and get as much information as possible from letters, photos, and phone calls. If you have a friend in the area where the plane is based, have him take a quick look. It might save you a trip.

Sellers sometimes receive many calls from dreamers and "tire kickers" so may not seem too enthused until you seriously talk about travelling to see it or place a deposit. If you ever sell a plane, you will soon see what the problem is.

To make the most of your buying trip expenses, plan to see more than one plane on a trip if possible. If you decide to do this, be sure the sellers know it and ask for their "rock bottom" price. Also ask them if there is anything you should know about the plane before you invest in the expense of a trip. If they are honest, it may save both of you time and money.

If you have pretty well decided beforehand to buy a certain plane at an agreed price, take a cashier's check with you to avoid a delay in transferring ownership and the trip home. If you want the plane, but are not sure how much you will have to pay, make out the cashier's check for the lowest amount you think the seller might take. If he won't sell at that price, make up the difference with cash or a personal check. The signed cashier's check can be an effective bargaining tool.

Don't fail to have a title search made before you take possession. Most of these older planes have passed through several owners, any one of whom could have had a lien placed on it. You might find you aren't the sole legal owner after all. This search should be done before you travel to see a potential purchase or you will be tempted to omit it so that you can buy the plane quickly and fly it home.

Inspecting The Plane

Your best inspection aid is your own A&P mechanic. Unfortunately there are times when you must rely on one you don't know. As a minimum, the engine compression should be checked and probably a fabric punch test should be performed. Problems in either area can run into hundreds or even thousands of dollars! The few dollars a mechanic charges will be well spent for the piece of mind and possible bargaining tool they will provide.

Fabric tests should always be performed on planes covered with cotton. "Lifetime" fabrics such as Ceconite, Eonnex, or Stits Polyfiber don't deteriorate as fast,but do require occasional replacement. Many owners replace these fabrics every few years on planes tied down outside. Even for planes that are always hangared, an occasional re-covering allows a thorough inspection of the underlying structure. This is especially important if the construction material is wood.

When checking fabric condition, remember that it deteriorates the fastest on the top surfaces where the sun hits. Also fabric deteriorates faster under dark colored paint than light.

There are two types of fabric testers, Maule and Saybolt. The Maule tester has a spring loaded plunger with an end shaped like this: 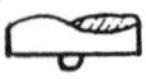 Pressing this end into the fabric will result in a numerical reading on the plunger when the small projection punches through. If the fabric supports a passing rating before it punches through, it is not necessary to make a hole in the fabric at all. The Saybolt tester has a spring loaded plunger with an end like this:  This will always make a hole in the fabric.

The reading it gives is on a color scale (green, yellow, orange, and red) depending on how much resistance the fabric gives to the penetrating plunger.

You can't expect a seller to agree to punch holes in his airplane unless you are serious about buying. One method of avoiding this is to check everything else and negotiate a price "providing the fabric is good". If it passes, you buy the plane at the agreed price, but if it doesn't he keeps the plane and patches the holes himself. A seller who won't agree to this probably has something to hide.

If fabric is applied incorrectly, it can warp the underlying structure. Usually this is caused by overshrinking the fabric. This will drastically affect the flying qualities of the plane and may even weaken it. If there are wrinkles anywhere in the fabric, check the underlying tubes for bending due to a hard landing or other overstress.

The tubing cluster just above the tailwheel should be carefully checked, especially on planes tied down outside. This is usually the lowest point in the fuselage so water tends to collect there. Repairs in this area are difficult and expensive. Rusted tubing can usually be detected by poking with an ice pick. The lower fuselage tubes should be checked the same way. It should be done from the outside, through the fabric, because water sometimes is trapped there and it is impossible to see from inside the fuselage. Dirt also holds moisture and hides rust damage. If the inside of the aircraft is especially dirty, scrape away the dirt to check for tubing corrosion and bad fabric. On planes tied down outside with the elevators up, be sure to check the leading edge tube of the elevator with the ice pick.

The tailwheel requires a close inspection. Check the condition of the tire or hard rubber wheel. Excessive internal wear of the steering system will result in a "shimmy" that often gets worse after lubrication. Replacement of the whole assembly can be surprisingly expensive.

Flight control cables should be checked for broken strands by sliding a cloth along them and feeling for snags. Do not use your bare hands or a broken strand will become painfully obvious!

Trim systems are often overlooked by potential buyers. Check that the handle in the cockpit operates the trim smoothly throughout its entire range. On aircraft with screw-jacks to lower and raise the leading edge of the horizontal stabilizer, excessive wear may be present due to the grinding action of dirt and sand trapped in the lubricating grease. Grasping the leading edge of the horizontal stabilizer and lifting and pressing down will show any sloppiness in the system. If the trim system uses a tab on the trailing edge of the elevator, insure that it is secure and operates smoothly.

Because of the potentially high cost of repairs to an engine a large part of the prepurchase inspection should be devoted to it. Excessive oil consumption, a mag drop that is out of limits, oil/fuel leaks, or low compression all merit further investigation by an A&P mechanic. Problems should be cleared up or the purchase price reduced.

Compression checks should be conducted while the engine is warm. Many people just pull the propeller through a few compression cycles and feel the resistance. Although this may uncover a gross loss of compression, most buyers prefer a more accurate test.

There are two types of cylinder compression testers, the plug type similar to automotive testers and the more accurate "differential pressure" tester. The first type is screwed into a sparkplug hole, the engine is rotated a few revolutions, and a reading is obtained from the protruding shaft much like a tire pressure gauge. This type has the advantage of being self contained. No compressed air is required.

The differential pressure type requires that compressed air be pumped into the cylinder with the piston at top dead center and both valves closed. The supplied air pressure is regulated to 80 psi. Any drop in pressure seen on the second gauge is due to leakage past rings and valves. A cylinder that holds significantly less pressure than the rest or cylinders that hold less than 50 psi should arouse suspicion. Most engines in top shape hold 70 psi or more.

One advantage of this type of test is its ability to help determine which engine component is causing the problem. If, when the test pressure is applied, air is heard escaping through the exhaust stack, the exhaust valve is probably not seating. This may mean the valve is burned, worn, or that a particle of foreign material such as carbon has lodged between the valve and its seat. If the air is escaping through the carburetor, the intake valve is the problem. Trouble with the rings can often be detected by listening through the oil filler tube.

A more positive means to determine whether rings or valves are bad is to squirt some engine oil into the cylinders and repeat the compression check. If the compression is appreciably improved, the rings were the cause. The oil will seal them temporarily.

However you do it, be sure to check the engine compression to avoid some very expensive surprises.

Do not overlook propeller condition during your inspection. If it is wooden, check that the mounting bolts have not been over-tightened, crushing the wood fibers under the hub. If the finish has been broken or worn through, moisture or oil may have soaked into the wood. The metal tips should be secure and the drain holes in the tips clear. Metal props should be free of corrosion and large nicks.

You should obtain a list of all AD Notes that pertain to the type aircraft, engine, and equipment such as ELT's. An Authorized Inspector or most A&P mechanics can show you a list of them. The ones that apply should have been accomplished and properly signed off - not just with a statement of "all AD's complied with".

The paperwork of older planes can become a nightmare. Learn what is required and insist that it be supplied. Otherwise, you will have to pay to have it corrected at the next annual inspection done by a conscientious inspector.

This section was not intended to frighten you with all the potential problems of buying an older aircraft. Being aware of them will help you avoid them. An informed buyer will probably become a more satisfied customer.

Univair
Rt 3, Box 59
Aurora, CO 80011

Wag-Aero
Box 181
Lyons, WI 53148

Trade A Plane
Crossville, TN 38555

FLYING HOMEBUILT TAILDRAGGERS

It is difficult to discuss a topic as broad as flying homebuilt taildraggers. There are so many variations - overpowered, underpowered, high wing loading, low wing loading, short fuselages, long fuselages, etc. - that one can only give general guidelines. Some parts of this section may seem very negative because of the emphasis of potential difficulties. In fact, some homebuilts are easier to fly and handle on the ground than most factory built taildraggers.

Design Factors That Affect Flying Qualities

Geometry of Major Components

The general layout of aerodynamic surfaces and landing gear play a powerful part in how easy a plane will be to fly and land. Homebuilts tend to be small. As a result they are "close coupled". This means the tail surfaces are not far from the CG and the tailwheel is close to the main gear.

An aircraft with a short fuselage tends to be lightly damped in yaw, requiring a lot of work with the rudder in turbulence. The lighter the plane is, the more this is exaggerated.

Control forces on small homebuilts are often very light. This is due to their small size and weight, the size of the surfaces, and a design philosophy that promotes a "sporty" feel. Sometimes this is a delightful characteristic. If carried to extremes, it can lead to overcontrolling and a control feel very foreign to the pilot flying the type the first time.

Unfortunately, this is most prevalent during takeoff when the propwash over the tail makes the rudder and elevator more sensitive. The problem is made worse by occurring before the new pilot has a chance to get the feel of the light forces in the air.

With a tailwheel that is close to the main gear, tailwheel steering may be sensitive. A little bit has a large effect. This is seen on some planes by a quick "S-ing" during takeoff and landing.

The width of the main gear is also important to the ease of maintaining control. In general, the wider the main gear is, the better. The major benefit is the extra margin available before the CG gets outside the main gear track which will end in a ground loop. There is also a side benefit of more effective steering with brakes.

Three Point Attitude

Ideally, the attitude of a tailwheel aircraft at rest on the ground will coincide with the stall angle of attack in ground effect at touchdown speed. Just as the plane reaches that attitude it will stop flying.

Often propeller clearance determines the three point attitude. It may cause the attitude to be more nose high than the ideal. This type of plane would then need to be wheel landed and not brought to a stall during landing.

If, in an attempt to improve visibility over the nose on the ground, a designer makes the attitude too flat, the plane will still be flying when three point landed and tend to bounce back airborne.

This design also will not allow the shortest landing roll possible because it is not flying at its minimum speed at touchdown. If brought to a stall during landing, it will touch tailwheel first.

Cockpit Visibility

If there is limited forward visibility on the ground, constant "S-ing" will be needed while taxiing. During takeoff, pick the tail up as soon as you have good directional control and if prop clearance allows.

Poor visibility while landing is another common problem with many homebuilts. You can fly a curving final or in a slight sideslip on final to help see runway alignment. In the flare, you should look out both sides of the aircraft for lineup. Looking out just one tends to cause a small sideslip as you subconsciously push a small amount of rudder to help see the alignment better. The touchdown attitude is perceived from peripheral cues. One good place to practice landing this way is in the back seat of a Piper Cub or similar plane. Noticing where the horizon intersects some part of the aircraft structure while in the three point attitude before takeoff will help you recognize the landing attitude later.

When flying up and away most monoplanes are in an attitude that allows reasonable forward visibility. A biplane often blocks large forward segments of the sky from view. It is misleading because often the view to the sides and directly above is so unobstructed you don't realise the important areas ahead are blocked. Clearing is especially important while descending or turning.

Wingloading

Wingloading, the ratio of weight to wing area, determines several characteristics of an aircraft - stall speed, turbulence response, reaction to wind gusts during landing, and size of change in angle of attack necessary to change the flightpath.

A light wingloading (little weight per wing area) tends to produce a slow stall speed and thus slow takeoff and landing speeds. This allows short takeoff and landing distances. A light wingloading gives good climb performance and low drag at low speeds.

A light wingloading is a benefit in most circumstances. Unfortunately there is a price to pay; turbulence will really bounce the plane around and wind gusts in the flare can be bothersome. The only way to alleviate the turbulence effect is to slow down.

A wind gust occurring when you are near the flare can pop you up several feet into the air. Then, when it quits, you may find yourself quite high with very little airspeed. It may be enough of a problem to require a go around and another attempt. You must remain alert until the plane is stopped.

Two techniques that help are: use of less than full flaps and slightly higher final approach speed to a wheel landing. You may still get tossed back into the air but you will have more airspeed with which to recover.

Aircraft with a high wingloading (heavy weight per wing area) will not be as sensitive to wind gusts but have other problems. They have higher takeoff and landing speeds, more drag at low speeds, and may require a higher nose high attitude in the flare.

Higher touchdown speeds result in an exaggeration of sensitvity in ground handling during rollout. A loss of directional control will have more dire consequences at the faster speeds also.

The increased induced drag of a highly wingloaded aircraft means power will be needed on final or a steep final may result. Speed will tend to bleed off rapidly in the flare.

If a large attitude change is required to arrest a high sink rate, more judgemant is required to make a power off landing. Carrying a little power throughout the flare will ease the landing problem somewhat.

Glide Ratio

The "dragginess" of an aircraft affects its ease of landing. If it has a high glide ratio like a sailplane, it will tend to float during landing. If it is too high it may require a spoiler to dump lift and get weight on the wheels. Most homebuilts aren't this extreme but they may be difficult to slow from cruising speed and require a flat final approach to keep speed in control.

Low glide ratios result from drag producing projections in the airstream or high induced drag at high angles of attack. Low glide ratios produce steep final approaches unless power is carried. As speed bleeds off rapidly in the flare, there is less time to "feel for the runway". Small biplanes are well known for this tendency. A little power throughout the flare often helps. Many pilots prefer to wheel land aircraft with this characteristic.

Flaps

Flaps on homebuilts range from ineffective to very powerful. Since they produce both lift and drag, their use can minimize other problems. In smooth air, nearly all types of aircraft benefit from additional lift, allowing slower and shorter landings. Usually the pitch attitude of a plane will be more nose down with flaps extended. This helps forward visibility on final.

In gusty winds lightly wingloaded aircraft will be less affected by the gusts if flaps aren't used. Flaps usually increase ground effect which may cause the plane to float excessively before touching down. Check the effect of sideslips on the plane with fully deployed flaps before attempting to use them in a crosswind. Sometimes they disrupt the flow over the tail, causing a pitch down motion.

Tailwheel Mechanism

The installation of the tailwheel often differs even from aircraft built to the same plans. Some designers leave a lot to the builder's imagination as to what tailwheel to use and how to mount it. Even identical mechanisms may be adjusted, lubricated, or worn differently.

Some builders leave tailwheel springs loose. This may keep the steering from being too sensitive. It will limit tailwheel excessive deflection during a crosswind landing. Other builders keep all slack out of the springs for accurate control. This may be necessary if the rudder is blanked when the aircraft is in the three point attitude. If the springs are tight and one becomes detached, the wheel will be cocked, probably causing a loss of control.

To minimize steering sensitivity, make very small rudder pedal deflections at high speed.

Brakes

The range of braking ability of homebuilts is from minimal, through "just right", to too much. Sometimes low braking power is intentional if, due to little weight on the tail, there is a tendency to nose over. If this is the case, just enough brakes to hold the plane for an engine runup check will be fine. If the aircraft doesn't require a lot of runway for takeoff and landing, weak brakes may be acceptable.

Brakes may be useful to steer during rollout if they are properly designed with the correct stopping ability and enough pedal throw so they can be controlled easily. If too strong or not easily controlled they can cause a loss of control, directly or by starting a swerve that results in a PIO ending off the runway.

Your First Flight

Let's look at how you might prepare for your first flight in a homebuilt taildragger. This discussion will assume this isn't the first flight of the airplane, but just the first flight you will fly.

Any pilot planning to fly a tailwheel homebuilt should have considerable experience, including tailwheel time. This is especially true if it is a single seater. Not all of them are difficult to fly - just different than modern training aircraft. A few hours of pattern work in a couple of different standard tailwheel aircraft will go a long way towards preparing you.

Try to get some recent practice in a two place aircraft that has the closest general layout and characteristcs of the homebuilt that you can find. If there is a two place version of the homebuilt available, that would be the best choice. Even if you aren't on the controls all the time, you will have some idea of how the takeoff and landing should look.

If there is no two place available, talk to others about how it flies and things to watch for. If possible, talk to someone who has flown the specific aircraft you will fly. Due to variations in workmanship and lack of completeness of some plans, not all examples of one type of homebuilt will flly the same. It is helpful if he can recommend a standard aircraft that flies somewhat like the one you will fly.

If no one can suggest a plane, the following list contains planes that may be useful in gaining practice. They weren't chosen because they are the most difficult to fly, but because they have characteristics often found in homebuilts.

Luscombe
Swift
Pitts Special (2 seat)
Piper PA17
Taylorcraft BC12
American Yankee

If you can find one with the same type of control (stick or wheel), about the same wingloading, and reasonably close control forces, you will get the most benefit.

The American Yankee, although not a taildragger, was included in the list because its flying characteristics are similar to many homebuilt aircraft.

There are two schools of thought as to how much high speed taxiing to do before actually flying. One group says that you are asking for trouble to accelerate up to near liftoff speed, chop the power, and slow back down to a stop. The other side says it is better to see handling characteristics at lower speeds than to wait until it is "all or nothing" in a landing.

One factor of importance in determining how much to high speed taxi is pilot experience. If you have flown other aircraft of the same general type before, feel confident in several other types of taildraggers, and have considered other factors such as runway width and type, you might decide against much taxiing.

The conservative approach would be to progress to higher and higher speeds slowly. On the first run, just get up to a speed about a third of the takeoff speed, ease the throttle back, and stop. If all went well, try a faster speed - say two thirds of takeoff speed - and then return to a stop. Runway length will play a part in how fast you ultimately will want to try without actually flying.

Do everything smoothly to keep control. When the throttle is retarded, control effectiveness will change when propwash stops flowing over the tail surfaces. Also gyroscopic effects will change with prop RPM and as you pick up the tail or set it down. Make changes slowly to minimize these effects.

ORDER FORM

Please send ________ copies of

TAMING THE TAILDRAGGER

@ $7.95 each - - - - - - - - $ __________

Plus Postage/Handling 2. 00

Check/Money Order Enclosed $ __________

NAME ____________________________________

STREET __________________________________

CITY ____________________________________

STATE ____________________ ZIP __________

Mail to : Goldstar Enterprises
PO Box 363
Williamsville, NY 14221